Chemistry and Medical Debate

To Doug Le Barr
a long time student
Allen Debus
Denver
November 2001

CHEMISTRY AND MEDICAL DEBATE

van Helmont to Boerhaave

ALLEN G. DEBUS

Morris Fishbein Professor Emeritus
of the History of Science and Medicine
The University of Chicago

Science History Publications / USA
2001

First published in the United States of America
by Science History Publications / USA
a division of Watson Publishing International
P.O. Box 493, Canton, MA 02021-0493

Library of Congress Cataloging-in-Publication Data

Debus, Allen G.
Chemistry and medical debate : van Helmont to Boerhaave / Allen G. Debus.
p. cm.
Includes bibliographical references and index.
ISBN 0-88135-285-3 (alk. paper)
1. Medicine–History–17th century. 2. Medicine–History–18th century.
3. Chemistry–History–17th century. 4. Chemistry–History–18th century.
5. Iatrophysical school. I. Title.

R148.D43 2001
610'.9–dc21

2001020100

Designed and typeset by
Publishers' Design and Production Services, Inc.

Manufactured in the U.S.A.

This book is dedicated to the many
Students, both Graduate and Undergraduate,
who took my courses and seminars at the
University of Chicago during the years 1961 to 1996

Contents

List of Illustrations

Preface

The present book owes its existence to a number of factors. More than fifty years ago an interest in both history and the sciences led me to a degree in chemistry with many courses in history at Northwestern University. From there I went to Indiana University to take a master's degree in history. My advisor there was John J. Murray, who encouraged me to do my research on Robert Boyle and seventeenth-century chemistry. Later I spent five years at Abbott Laboratories as a chemist in research and development. There I became acutely aware of the close connection between chemistry and medicine in current research. My interest in the origins of this connection led me to leave Abbott Laboratories to return to graduate work, this time at Harvard University. In 1956 the history of science was centered on the development of the physical sciences in the period from Copernicus to Isaac Newton. The history of chemistry was seldom mentioned and as far as the history of medicine was concerned, it seemed largely divorced from then current interests in the history of science and there was little interest in seeking any connections. Among historians of medicine few were concerned with the work of historians of science . . . in short, the fields seemed to be quite separate.

At that time the main emphasis of historians of science was on the sixteenth and seventeenth centuries. The new science that developed at that time was pictured as experimental, mechanical, and inspired by mathematics and quantification. The founder of twentieth-century history of science, George Sarton, had been trained as a mathematician and he had been strongly influenced by August Comte. With those interests it is not surprising that Sarton presented a positivistic approach to the development of the sciences that to a large extent excluded alchemy and the early history of chemistry. Nor did he consider medicine to be the equal of the mathematically based sciences. Forty-five years ago the aspiring scholar in the field was encouraged to work on some subject related to astronomy or the physics of motion in the period of the scientific revolution. If, indeed, he had an interest in the biological sciences, he might turn to some aspect related to the discovery of the circulation of the blood by William Harvey, but an interest in non-scientific topics (from our point of view) was not encouraged. Historians of science at that time seldom ventured into the relationship of the sciences with societal matters. It was essentially an "internalist" discipline emphasizing the technical aspects of the subject.

My own interests then as now centered on the same chronological period of interest to those working on problems considered mainstream in the Scientific Revolution, but instead of Copernicus, I became interested in his contemporary, Paracelsus, and instead of Galileo, I turned to the chemist and physician, van Helmont. If Copernicus and Galileo founded a new astronomy and physics of motion, Paracelsus and van Helmont sought a new understanding of the world through chemistry and chemical analogies that was to profoundly affect medicine as well as the sciences. And although the standard interpretation of the Scientific Revolution postulated the rise of a mechanical philosophy concurrently with the decline of Aristotelian natural philosophy and Galenic medicine, I saw a third factor leading to a new science and medicine, the attempt of the Paracelsians to establish their chemical philosophy as the basis of a new understanding of nature, both macrocosm and microcosm. The addition of this factor added to the complexity of the Scientific Revolution since the Aristotelians and Galenists were not opposing a new astronomy and physics alone, but the chemists as well. In a series of books and papers including *The English Paracelsians* (1965), *The Chemical Philosophy: Paracelsian Science and Medicine in the Sixteenth and Seventeenth Centuries* (1977), *Man and Nature in the Renaissance* (1978), and *The French Paracelsians: The Chemical Challenge to Medical and Scientific Traditions in Early Modern France* (1991), I sought to describe the chemical philosophy of these iatrochemists, their debates with the Aristotelians and Galenists, and then to indicate the role they played in the Scientific Revolution. The present book goes on to discuss the late seventeenth and early eighteenth centuries to show the rise of a new group of medical physicists (iatrophysicists) who recognized the important role played by mathematics and physics in the development of the new science and hoped to apply the same methods to medicine. Like the Aristotelians before them, these mechanist physicians found themselves at odds with the chemical physicians. Here we see that the debate of the chemists and the mechanists in medicine continued well into the eighteenth century. I have attempted to present a survey of the interaction of chemistry and medicine in these two centuries. The material available for such a study is enormous and in no way do I claim to have covered it all. Still, I hope to have touched on many of the major figures and topics in this often acrid development.

Although I have here a special interest in the late seventeenth and early eighteenth centuries, I have opened the volume with a short survey of the work of Paracelsus and its impact on the medical world of the sixteenth century (Chapter 1) followed by a discussion of Jean Baptiste van Helmont and the immediate reaction to his opera (Chapter 2). Some of this material is new, but much of it is derived from my earlier publications. In any case, I think that it is essential for an understanding of later developments relating to chemistry and medicine in this period. The third chapter takes up the medico-chemical views of Franciscus de la Boë Sylvius, Thomas Willis, and Robert Boyle prior to discussing some of the literature related to the ill-fated attempt to establish a society of chemical physicians in London in 1665. The following chapter describes the strong interest in an acid-alkali dichotomy in medicine and chemistry in the late seventeenth century. Here I have emphasized the work of Otto Tachenius, who claimed that the ancient physician, Hippocrates, was a chemist.

With the rise of a new physics based on mathematics, there were many who hoped to apply Newtonian methods as the basis of a new medicine (Chapter 5). Here I center my remarks on Archibald Pitcairne and Georgio Baglivi before going on to the early eighteenth-century debate on digestion in which Raymond Vieussens and Jean Astruc were the chief proponents of a chemical interpretation in opposition to the mechanist, Philippe Hecquet. A corollary to this development is offered in a discussion of the late development of chemical medicine in Spain.

In Chapter 6, I turn first to the histories of medicine penned by Daniel LeClerc and John Freind to indicate how personal convictions could affect their interpretations of events. This is followed with a discussion of the mechanistic medicine of Hermann Boerhaave and the animistic medicine of Georg Ernst Stahl. Both were enthusiastic chemists who influenced the development of this science in the course of the eighteenth century: the first through his very popular textbook and the second through his statement of the phlogiston theory. However, neither one approved of the use of chemistry or chemical explanations in the study of the human body. This is followed by a discussion of the reaction to Stahl's work by Friedrich Hoffmann and Gottfried Wilhelm von Leibniz. A final chapter (7) reviews the period from Paracelsus to the eighteenth century prior to discussing the work of several mid-eighteenth century authors and then briefly concludes with the impact of the new chemistry of Lavoisier and his colleagues at the end of the century.

In the course of this research I have become indebted to many individuals and institutions. In addition to the holdings of the Department of Special Collections at the University of Chicago, I have made many trips to the University of Wisconsin at Madison, where the extraordinary Duveen and Cole Collections of early chemical works are housed. There John Neu has always been most helpful in locating rare items for me. A third port of call has been the National Library of Medicine in Bethesda, which is also distinguished by its extensive collection of early medical and chemical works. The NLM is surely one of the most reader-friendly libraries I know. Beyond the United States, libraries I have always turned to are the British Library and the Wellcome Historical Medical Library in London. I am particularly indebted to the latter for facilitating my research as a visiting scholar during the summer of 1994 and the spring of 1997. I am grateful to the National Institutes of Health for research support during the early stages of this project (Research Grant RO1-LM-5407-01-02-03-04, 1992–1996).

Invitations to a number of meetings and universities made it possible to discuss aspects of my current research with other scholars. An invitation from the Istituto Italiano per gli Studi Filosofici in Naples (March, 1990) gave me the opportunity to review my research on the Paracelsian tradition in a series of four lectures later published as *Paracelso e la Tradizione Paracelsiana.* Later that year I gave a course on the history of alchemy and chemical medicine from Paracelsus to Boerhaave and Stahl at the Instituto de Química of the Universidade de São Paulo in Brazil. I returned there five years later to discuss my current research on the chemical debates of the seventeenth century at a meeting of the Brazilian Society of the History of Science. The five hundredth anniversary of the birth of Paracelsus gave me the opportunity to review the contributions of Paracelsus while I acted as a consultant to a book display at the National Library of Medicine. Also in 1993 I spoke on "Paracelsianism and the Diffusion of the Chemical Philosophy in Early Modern Europe" at a meeting in Glasgow celebrating that anniversary. Professor A. J. A. Gouveia invited me to speak at the University of Coimbra and the Academy of Sciences in Lisbon in 1995 and it was at that University that I spoke on "The Rise and Fall of Chemical Physiology in the Seventeenth Century." And Michael T. Walton joined with me to organize a series of sessions largely on Paracelsian topics at the St. Louis meeting of the Sixteenth Century Studies Conference in 1996. I spoke there on "Paracelsus and the

Delayed Scientific Revolution in Spain: A Legacy of Philip II." These papers and others are reflected in the pages of this book.

It would be impossible to close this preface without mentioning Professor I. B. Cohen, who permitted me to pursue my research at Harvard University in an area that was then somewhat divorced from mainstream studies. And I will never be able to repay my debt to Walter Pagel, a friend whose work inspired many scholars who work in this area today. The thirty-five years I taught at the University of Chicago brought me in contact with a remarkable number of talented students, many of whom are now well known in the international community of scholars. The discussions I had with them in my seminars were stimulating and added materially to the formation of my own views. Here let me refer only to my former student, Karen Parshall, now Professor of History and Mathematics at the University of Virginia, and my current student, Ku-ming Chang, for their constant interest and help. I wish also to thank Mrs. Elizabeth Bitoy, my secretary for twenty years, who typed the first four chapters of this manuscript along with many other papers in addition to her other work prior to her retirement in 1998.

And finally, one last word of gratitude to my wife, Bruni, who has been a constant source of support and who has had to put up with my trips to libraries and to meetings both here and abroad, often at inconvenient times.

Allen G. Debus
Deerfield, Illinois

CHAPTER 1

The Renaissance Chemical Revolution

[Chemistry] openeth unto us so many secretes of nature, and preparations of herbes, beastes and mineralls hetherto unknowen, and sheweth the uses almost of all things, which were hidden and laid up on the bosome of Nature, that they shew themselves unkinde toward man, that would have this art buried.[1]

Joseph Quercertanus (Duchesne) as translated by John Hester, 1591

In the course of the sixteenth and early seventeenth centuries, the traditional natural philosophy and medicine of the schools had been rocked by a series of publications and discoveries that had led to deep divisions among scholars. To be sure, the Copernican statement of the heliostatic universe was of importance for the development of science as we know it today, but the debate over chemistry then was of equal or greater importance because it seemed to threaten traditional beliefs in medicine, natural philosophy, and cosmology–the whole range of the sciences and medi-

[1]Joseph Quercetanus (Joseph Duchesne), *A Breefe Aunswere . . . to the exposition of Iacobus Aubertus Vindonis, concerning the original, and causes of Metalles. Set foorth against Chimistes. Another exquisite and plaine Treatise of the same Josephus, concerning the Spagericall preparations, and use of minerall, animall, and vegitable secretes, not heertofore knowne of many.* By Iohn Hester, practitioner in the Spagericall Arte (London, 1591), pp. 2–3.

cine from our present viewpoint.[2] A subject that had earlier been confined largely to the manipulations of alchemical charlatans, metallurgical assayers, or those apothecaries who prepared the limited number of metallic- or mineral-based remedies known to the ancients, chemistry suddenly found proponents who sought to make it the new key to an understanding of both microcosm and macrocosm. This claim was not one that was to be readily accepted by conservative physicians or natural philosophers who found the works of the ancients more compatible with their beliefs.

RENAISSANCE HUMANISM

While there is no need to discuss the background of all the seventeenth-century debates in detail, it may be helpful to indicate the origins of some of the conflicts that were to affect the sciences. There have probably been few times in history–if any–in which the study of nature was completely static. There have always been those who question or disagree with authorities. This certainly was the case in the Middle Ages, and the Renaissance was no different.

The so-called Twelfth Century Renaissance had produced a large number of Latin translations of ancient classics and many of these had been made from Arabic texts which had in turn been made from Greek originals. The new Latin translations often carried over old errors and introduced new ones as well. As early as 1267 Roger Bacon (1214–1292) complained that

> . . . so great is the perverseness, crudity, and terrible difficulty in the translated works of Aristotle that no one can understand them, but each one contradicts another, and false statements are found again, as is clear from a comparison of the different translators and the texts of the different languages.[3]

---

[2]The present introductory chapter is based on a series of my earlier books and papers which will be cited throughout. For a general overview see Allen G. Debus, *Man and Nature in the Renaissance* (Cambridge University Press, 1978); for a discussion of the Paracelsian tradition, see Allen G. Debus, *The Chemical Philosophy: Paracelsian Science and Medicine in the Sixteenth and Seventeenth Centuries* (2 vols., New York: Science History Publications, 1977) and Allen G. Debus, *Paracelso e la Tradizione Paracelsiana* (Napoli: La Città del Sole, 1996).

[3]Roger Bacon, *The Opus Majus of Roger Bacon*, trans. Robert Belle Burke [2 vols., New York: Russell and Russell, Inc., 1962 (reprint of 1928 edition)], *1*, p. 77.

With the Renaissance of the fifteenth century, a new effort was made to find copies of ancient texts in their original Greek so that more accurate–and more elegant–translations could be made. The problem of reading Greek had been at first formidable, but with a series of Church Councils seeking to bridge the gap between the Christianity of East and West, some Byzantine scholars settled in Italy and taught their language to eager would-be translators. At first literary classics were sought–history, rhetoric, and poetry–but by the end of the century there was an increasing interest in natural philosophy and medicine. Many of the most dramatic discoveries related to astronomy. George von Peurbach's (1423–1461) realization of the need for an accurate Greek text of the *Almagest* led to Johannes Müller's (Regiomontanus) (1436–1476) discovery of astronomical manuscripts in Italy and his subsequent *Epitome* (1496) of Ptolemy's great work. Its appearance when Copernicus (1473–1543) was a student was surely an important influence on him.

The recovery of the medical treatise of Celsus (fl. 25 A.D.) offered to humanists a technical treatise in Latin penned at a time of stylistic elegance. But more significant was the recovery of Galenic works previously unknown. Although many ancient medical texts had been translated in the twelfth and thirteenth centuries, some had only been known in fragmented form and in any case these translations had been cast in a form of Latin considered barbarous in the fifteenth century. The search for new texts and the preparation of accurate translations became a major pursuit of these medical humanists.[4] In England Thomas Linacre (1460–1524) translated many of the works of Galen and other ancient physicians, as did his younger Parisian counterpart, Jacques Dubois (1478–1555). A figure of special importance was the Parisian Professor of Medicine, Johannes Guinter of Andernach (c. 1497–1574). A tireless translator, Guinter produced editions of Paul of Aegina, Caelius Aurelianus, Oribasius, and Alexander of Tralles. However, he, like other early sixteenth-century physicians, gave a primary role to Galen. Two of his most gifted students, Andreas Vesalius (1514) and Michael Servetus (ca. 1511–1553) reflected that emphasis, the first by maintaining Galenic explanations even when they seemed to contradict observation, and the second by openly confessing that

[4]On Renaissance medical humanism see the collection of papers edited by A. R. Wear, K. French, and I. M. Lonie, *The Medical Renaissance of the Sixteenth Century* (Cambridge University Press, 1985).

> In our happy age [Galen] once shamefully misunderstood is reborn and reestablished himself to shine in his former lustre: so that like one returning home he has delivered the citadel which had been held by the forces of the Arabs, and he has cleansed those things which had been bespattered by the sordid corruption of the barbarians.[5]

For these medical humanists Galen was clearly the "Prince of Physicians." It seemed that medicine had now been reestablished on a solid foundation.

But the research of the past forty years has confirmed the fact that the sciences were affected by texts other than those of the ancient physicians and natural philosophers. The fifteenth century saw a renewed interest in the Platonic dialogues and when the *Corpus Hermeticum* was recovered (ca. 1460), it seemed to offer a new avenue to truth.[6] A relatively short collection of tracts originally written in the third or fourth centuries A.D., it reflected Platonic, neo-Platonic, and mystical religious currents common to that period. The *Pymander*, perhaps the best-known text of the collection, presented a Creation account similar in many ways to the first chapter of Genesis. However, instead of leaving man devoid of divine knowledge after the Fall, he was here left with the power to act in nature on his own. Such a man was a true natural magician who might seek those treasures hidden from mankind at the time of the expulsion from the Garden of Eden. As in Plato's *Timaeus*, these hermetic treatises relied on the analogy of the macrocosm and the microcosm, and it was felt that a magus, a true natural magician, would be able to understand man, the microcosm, through his study of the macrocosm because the former was a perfect representation of the latter. Physicians were to find this a new key to their work. No less appealing was the fact that this call for new observations in nature could be seen as an act of devotion. Christians should study not only Holy Scripture, but also the book of Nature, which was without doubt a second work of divine revelation.

---

[5]Michael Servetus, *A Complete Account of Syrups Carefully Refined According to the Judgment of Galen . . .* in *A Translation of his Geographical, Medical and Astrological Writings with Introduction and Notes*, trans. Charles Donald O'Malley (Philadelphia: American Philosophical Society, 1953), p. 60.

[6]Recent interest in the Hermetic background to modern science may be traced in large part to the numerous works by Dame Frances A. Yates and to the controversies resulting from them. See in particular her *Giordano Bruno and the Hermetic Tradition* (Chicago: The University of Chicago Press, 1964), pp. 1–116. See also the works of Cesare Vasoli and Paolo Rossi.

When translated by Marsilio Ficino (1433–1499) the *Corpus Hermeticum* was accepted not as a product of late antiquity, but as a true work of the Egyptian Prince, Hermes Trismegistus, who was thought to have lived at about the time of Moses or Abraham. Saint Augustine and other Church Fathers had accepted the works as genuine . . . why shouldn't the Renaissance humanists do the same? Here then was a gentile revelation that offered greater hope to mankind than the Old Testament. Only later with the research of Isaac Casaubon (1559–1614) was the late date of these texts established in 1617, but throughout the seventeenth century hermetic philosophers ignored the evidence.

The tracts of the *Corpus Hermeticum* were for the most part new to western Europe. Only one of them, the *Asclepius*, had been known in Latin translation during the medieval period. And yet, the name Hermes was well known through the works of the Church Fathers of late antiquity. Hermes was also one of the great figures of alchemical tradition. Alchemy was indeed the hermetic science. To be sure, alchemy implied the transmutation of the base metals, but more important was the separation by chemical means of the pure essence of a substance from its impurities. Through such processes the true divine signatures hidden from man at the time of the Fall might be recovered. In this fashion one would learn more of the Creator while recovering His priceless gifts though such endeavors. The result of these labors would frequently be products of near miraculous medicinal value and it was the hope of these chemists that the crude substances described in herbals would be replaced by pure distilled essences. Distillation had been applied to medicine earlier through the analysis of spa waters, but by the beginning of the sixteenth century distillation as a means of isolating medical "virtues" was being widely recommended.

In short, by 1500 the impact of the newly recovered texts was leading in two directions. On the one hand the natural philosophers and physicians of the schools developed an increased respect for Aristotle and other ancient authorities. Galen, in particular, reached a new level of importance as the "Prince of Physicians." On the other hand, the recovery of the *Corpus Hermeticum* and other more mystical texts placed an emphasis on natural magic and the relationship of man to the macrocosm, and sought divine truths in the study of nature. The first led to truth through traditional medicine and a reliance on mathematics and the physics of motion for our understanding of nature; the second led to a more mystical and religious basis of knowledge and turned to chemistry as a key to man and nature alike.

PARACELSUS AND CHEMICAL MEDICINE

Born in the small Swiss town of Einsiedeln, Philippus Aureolus Theophrastus Bombastus von Hohenheim (1493–1541)[7] may seem at first glance to have been an unlikely candidate to bring about any sort of major intellectual reform. And yet, some of his contemporaries were to give him the title, "Luther of Medicine," while he gave himself the name Paracelsus, meaning modestly that he was greater than the ancient Roman medical authority. In fact, this rural boy learned a great deal both from his surroundings and from those with whom he came in contact at an early age. The son of a physician, he was raised in Swiss mining towns. There he learned the rudiments of medical practice by observing his father. More specifically he learned something of metallurgical practices and of the occupational diseases afflicting miners–a subject on which he later wrote. But his father was also interested in alchemy and the boy could watch him as he carried out various operations at the family hearth.

Alchemy, the folk medicine of southern Germany and Switzerland, and the metallurgical practices of miners–these were all absorbed by the youth and were to be later reflected in his work. And when he left home at the age of fourteen to begin years of travel, he carried this knowledge with him. Paracelsus was later to write that he had seen much of Europe, the Near East, and North Africa, but little evidence can be advanced to support these

[7]The literature on Paracelsus is very extensive, but the most significant recent study is that of Walter Pagel, *Paracelsus: An Introduction to Philosophical Medicine in the Era of the Renaissance* (Basel: S. Karger, 1958; second revised edition, 1982). In 1993 a number a meetings were held to mark the five-hundredth anniversary of the birth of Paracelsus. The resultant conference volumes include *Das Werk-die Rezeption. Beiträge des /symposiums zum 500.Geburtstag von Theophrastus Bombastus von Hohenheim, genannt Paracelsus (1493–1541) an Universität Basel am 3. Und 4. Dezember 1993.* (Stuttgart: Steiner, 1995) [includes papers by Vivian Nutton, "Der Luther der Medizin: Ein paracelsisches Paradoxon"; Bernhard Dietrich Haage, "Paracelsus kein Luther der Medizin"]; *Un esame critico del pensiero di Paracelso, collocato nella sua dimensione storica. Atti del Convegno Internazionale su Paracelso, Roma 17–18 dicembre 1993, Goethe Institut, Roma* (Roma: Edizioni Paracelso, 1994) [includes papers by Wolf-Dieter Müller-Jahncke, Charles Webster, Piyo M. Rattansi, and Giancarlo Zanier]; *500 Jahre Paracelsus: Nachlese zum Jubiläumkongress* (Salzburger Beiträge zur Paracelsusforschung, 28) (Wien: Osterreischer Kunst- und Kulturverlag, 1995). In 1993 an exhibit at the National Library of Medicine in Bethesda, Maryland resulted in a brochure, *Paracelsus, Five Hundred Years: Three American Exhibits* (Bethesda, Maryland: Published by the Friends of the National Library of Medicine, Inc., for the Hahnemann University Library, The National Library of Medicine, and the Washington University School of Medicine, St. Louis, 1993) [includes Allen G. Debus, "Paracelsus and the Medical Revolution of the Renaissance: A 500th Anniversary Celebration"].

claims. It is only after 1520, when we find him back in central and southern Europe, that we can follow him as he moved from town to town writing and practicing medicine. At last he was called to Salzburg by the suffragan bishop Ernest of Wittelsbach, and died there at the early age of forty-eight in 1541.

In the course of his travels, Paracelsus had visited a number of universities and there is a possibility–certainly not proven–that he received an M.D. degree at Ferrara. However, he also served as a surgeon in the mercenary armies of the period, an unlikely post for a university-trained physician. Surely his practical experience in the field renewed his acquaintance with folk medicine and there is little question that his contact with academic medicine left him with a strong distaste for the medical establishment. For the most part Paracelsus preferred to associate with commoners, but in 1527 he had been called to Basel by the humanist publisher, Frobenius. At his home Paracelsus was in contact with a select circle of intellectual giants of the day, including Erasmus and Oecolampadius. Fortune seemed to smile on him when he was appointed city physician and professor of medicine. The latter was a city appointment, but it gave him the right to lecture at the university. Here he ran into trouble, for he had little respect for the academic medicine of his day. The story of his burning of the *Canon* of Avicenna on the St. John's Day bonfire has been told often, but his contemptuous address to the physicians bears repeating because it is so characteristic of his attitude.

> I am Theophrastus, and greater than those to whom you liken me; I am Theophrastus, and in addition I am *monarcha medicorum* and I can prove to you what you cannot prove. I will let Luther defend his cause and I will defend my cause, and I will defeat those of my colleagues who turn against me; and this I shall do with the help of the *arcana*. . . . It was not the constellations that made me a physician; God made me . . . I need not don a coat of mail or a buckler against you, for you are not learned or experienced enough to refute even one word of mine. I wish I could protect my bald head against the flies as effectively as I can defend my monarchy . . . I will not defend my monarchy with empty talk, but with *arcana*. And I do not take my medicines from the apothecaries; their shops are but foul sculleries, from which comes nothing but foul broths. As for you, you defend your kingdom with belly-crawling and flattery. How long do you think this will last? . . . Let me tell you this: every little hair on my neck knows more than you and all your scribes, and my shoe

> buckles are more learned than your Galen and Avicenna, and my beard has more experience than all your high colleges.[8]

In short order Paracelsus alienated the medical faculty of the University and the students who called him "Cacophrastus." Even his treatment of Frobenius was a failure because his patron died in October, 1527. Then, in the course of a dispute over a fee for his treatment of a church dignitary, Paracelsus fled Basel. He had lost all his support in the city.

Most of his works date from this late period when he was wandering through the towns of southern Germany and Switzerland. And yet, little of his work was published during his lifetime. His important *Der grossen Wundarznei* appeared in 1536 and 1537, but most of his early publications were popular prognostications. During his lifetime he was known less for what he had written than for his reputed skill as a physician.

All this was to change after mid-century. A series of cures using Paracelsian remedies resulted in the conversion of a number of Galenic physicians. Their enthusiasm led to a search for the manuscripts of Paracelsus and as these were recovered, they were published with notes and commentaries by these new disciples. Appearing first as individual works and then as small collections, the end result was the authoritative ten-volume opera prepared by Johannes Huser over the years 1589 to 1591.[9] Reprinted and translated into Latin in the seventeenth century, the Huser collection finally offered readers the medical and philosophical works of Paracelsus in a form that has remained standard down to our own century.

In these works one reads a bold challenge to the educational establishment. Reliance on the ancients in natural philosophy and medicine could no longer be tolerated. Further, the very method of learning was to be rejected. The study of logic and the search for truth through disputations built only on past knowledge must be rejected. How could any reform proceed if the object of the system was to protect traditional education? The English Paracelsian, Rychard Bostok, stated bluntly in 1585 that

---

[8]This passage from the *Paragranum* has been translated frequently. Here I give the translation in Paracelsus, *Selected Writings*, ed. with an introduction by Jolande Jacobi, trans. Norbert Guterman (New York: Pantheon Books, Bolligen Series XXVIII, 1951), pp. 79–80. For the original text see Paracelsus, *Sämtliche Werke*, ed. Karl Sudhoff and Wilhelm Matthiessen [15 vols., Munich and Berlin: R. Oldenbourg (vols. 6–9: O. W. Barth), 1922–1933], *8*, pp. 63–65.

[9]The essential guide to the publications of Paracelsus is Karl Sudhoff's *Bibliographia Paracelsica* (Berlin: Verlag Georg Reimer, 1894; reprinted Graz: Akademische Druck - u. Verlagsanstalt, 1958).

> in the scholes nothing may be received nor allowed that savoreth not of Aristotle, Gallen, Avicen, and other Ethnickes, whereby the yong beginners are either not acquainted with this doctrine, or els it is brought into hatred with them . . . likewise, the Galenists be so armed and defended by the protection and authoritie of Princes, that nothing can be allowed that they disalowe, and nothing may bee received that agreeth not with their pleasures and doctrine.[10]

This rule of the ancients must end–not only because it perpetuated the errors of the past, but because the ancient authorities were pagans. Their philosophy must be replaced with a Christian philosophy more appropriate to a world that was experiencing a religious reformation.

Here the Paracelsians turned from textual authority to nature. We must learn to read the great book of nature as we read the Bible itself since both represent divine Revelation. Peter Severinus (1542–1602), physician to Frederick II of Denmark, urged his readers to

> sell your land, your houses, your clothes and your jewelry; burn up your books. On the other hand, buy yourselves stout shoes, travel to the mountains, search the valleys, the deserts, the shores of the sea, and the deepest depressions of the earth; note with care the distinctions between animals, the differences of plants, the various kinds of minerals, the properties and mode of origin of everything that exists. Be not ashamed to study diligently the astronomy and terrestrial philosophy of the peasantry. Lastly, purchase coal, build furnaces, watch and operate with the fire without wearying. In this way and no other, you will arrive at a knowledge of things and their properties.[11]

There is little doubt that the Paracelsians applauded the gathering of knowledge through fresh observations in nature and in their laboratories.

[10] R. B. Esquire (Rychard Bostok), *The difference betwene the auncient Phisicke first taught by the godly forefathers, consisting in vnitie peace and concord: and the latter Phisicke proceeding from Idolaters, Ethnickes, and Heathen: as Gallen, and such other consisting in dualitie, discorde, and contrarietie . . .* (London: Robert Walley, 1585), sig. Fii^{v}. On Bostok see David Harley, "Rychard Bostok of Tandridge, Surrey (c. 1530–1605), M.P., Paracelsian propagandist and friend of John Dee," *Ambix*, *47* (2000), 29–36.

[11] Petrus Severinus, *Idea medicinae philosophicae* (3rd ed. Hagae Comitis: Adrian Vlacq, 1660), p. 39. The first edition appeared in 1571.

On this score there is little to separate them from the early mechanists. However, in other respects there are considerable differences. The modern scientist would argue that the application of mathematics to both physical and biological problems differentiates our science from that of the ancients. The Paracelsians thought otherwise. For them the works of Aristotle were too mathematical. Thus Petrus Severinus argued that the work of Aristotle was based upon mathematical laws and theorems. And, impressed with these texts, Galen had sought to mathematicize medicine–an endeavor that was doomed from the start.[12] But the words of Severinus are deceptive. For him mathematics meant the use of proof through logical progression as one finds in geometry. Indeed, the attack on deductive logic was common to much of the new science of the early modern period. For the Paracelsians other forms of mathematics were more certain. Some looked upon the more mystical use of numbers seen in neo-Platonic and neo-Pythagorean texts as an important path to truth: surely all referred willingly to the Biblical statement (Wisdom 11: 17) that God had created "all things in number, weight and measure." Here laboratory practice was accepted without the questionable referral to mathematical abstraction.[13]

But if our mathematical approach was abhorrent to most of these reformers, where did they turn? Surely Paracelsus and his immediate followers believed firmly in an absolute analogy of the macrocosm and the microcosm. In an early work, the *Paramirum* (ca. 1520), he had explained that

> the firmament is within man, the firmament with its great movements of bodily planets and stars which result in exaltations, conjunctions, oppositions and the like, as you call these phenomena as you understand them. Everything which astronomical theory has searched deeply and gravely by aspects, astronomical tables and so forth,–this self-same knowledge should be a lesson and teaching to you concerning the bodily firmament.[14]

This analogy was a true key to knowledge that offered to us the wondrous wisdom of the Creator.

---

[12] *Ibid.*, pp. 2, 3, 21.

[13] For a general discussion of mathematics in the Paracelsian texts see Allen G. Debus, "Mathematics and Nature in the Chemical Texts of the Renaissance," *Ambix*, *15* (1968), 1–28, 211.

[14] Paracelsus, *Volumen medicinae paramirum*, trans. and preface by Kurt F. Leidecker, *Supplement to Bulletin of the History of Medicine*, no. 11 (Baltimore: The John Hopkins Press, 1949), p. 36.

In time this cosmic analogy was to be discarded, but a second key to understanding became more important. This was chemistry. Here again Paracelsus could be cited. In the posthumously published *Philosophy to the Athenians* (1564) he had explained the process of Creation as that of chemical separation. God was no less than a divine alchemist. As the English Paracelsian, Thomas Tymme (d. 1620) argued,

> *Halchymie* should have concurrence and antiquitie with Theologie . . . [because Moses] . . . tells us that the *Spirit of God moued upon the water*: which was an indigested Chaos or masse created by God, with confused Earth mixture: yet, by his Halchymicall Extraction, Seperation, Sublimation, and Coniunction, . . . [the four elements, Earth, Water, Air and Fire, were produced].[15]

For the Paracelsians, Genesis I was looked upon as a scientific document that explained the origin of the elements. This point was important because ancient element theory was basic for an understanding of humoral medical theory. Because each of the four elements was associated with two of the four qualities (hot, cold, wet and dry)–and because these qualities could be altered–it was theoretically to be expected that the elements could be changed into one another. It was a complex system, but a potentially fragile one, since a rejection of even one of the four might result in the collapse of the entire edifice. Again it was the destiny of Paracelsus to initiate this attack.

The quotation above from Thomas Tymme correctly reflects the willingness of Paracelsus to use the traditional four elements. But at the same time he added salt to the time-honored sulphur-mercury theory of the metals, offering the *tria prima*, the three principles which also were endowed with elemental properties. He himself used the two systems without consistency and this introduced an uncertainty into the subject that lasted for more than a century. Still, for the working chemist the *tria prima* seemed to correspond to real substances–taste and solidity (salt), oil or inflammability (sulphur), and spirit (mercury). Discussing the problem of the elements, Bernard G. Penotus (ca. 1520 or 1530–1620) questioned his contemporary Galenist opponents:

[15]From the dedication to Sir Charles Blunt in Joseph Duchesne (Quercetanus), *The Practise of Chymicall, and Hermeticall Physicke, for the preseruation of health*, trans. Thomas Tymme (London: Thomas Creede, 1605), sig. A3^{r}.

> I say, every *Paracelsian*, which doth but onely carry coals unto the work, can shew you by eye three principles of *Theophrastus* Physick. Have you tasted the most sharp Salt, or the most sweet Oil, or the Balm, that most delicate liquor? All those being hidden in every thing that is created, you have not once perceived. The metalline spirits, in whom Physick doth consist, by no means can be found out, neither what force they have, or fellowship with mans nature, but onely by fire. . . .[16]

Element theory was only one aspect of macrocosmic concern. The Paracelsians turned to chemical processes for the explanation of natural phenomena.[17] Distillation was the model employed for rain, volcanic eruptions, and the origin of mountain streams. Indeed, the earth was viewed as a vast distillation flask with a fiery center that vaporized underground water deposits that then rose to the cool mountain tops before condensing in the form of mountain streams. A similar process resulted in lava flows. To be sure, the concept of a central fire was hardly new, but these explanations took on a more explicitly chemical appearance. Similarly, the Paracelsians maintained the vitalistic world of the ancients. The earth was a living being and the metals grew in veins much as a fetus grows in a woman. This was also an essentially alchemical viewpoint and the belief was widespread that, given time, the metallic ores developed into metals and that the base metals would perfect themselves by changing into silver or gold. The process was initially a sexual one originating from the union of an astral seed and the proper earthly matrix.[18]

Life was sustained in all things through the presence of a life spirit essential for both the organic and inorganic worlds. By the final decade of the sixteenth century this spirit was identified as an aerial niter or saltpeter in a tract

[16]Bernard G. Penotus, "Apologeticall Preface" to Paracelsus, *A hundred and fourteene Experiments and Cures of the famous Physitian Philippus Aureolus Theophrastus Paracelsus . . . Collected by Iohn Hester* (London: Vallentine Sims, 1596), sig. Ddd l^r.

[17]See the discussion of "Chemistry and the Geocosm" in Debus, *Chemical Philosophy, 1*, pp. 84–96.

[18]As described in Basil Valentine, *The Last Will and Testament of Basil Valentine, Monke of the Order of St. Bennet . . . To which is added Two Treatises, the First declaring his Manual Operations. The Second shewing things Natural and Supernatural . . .*, trans. J(ohn) W(ebster) (London: S. G. and B. G. for Edward Brewster, 1671), pp. 469–70 ("Preface"); p. 512 ("Of Things Natural and Supernatural").

ascribed to Paracelsus and printed in the Huser edition of his works.[19] This concept is one that rapidly gained currency among chemists, some of whom sought to isolate the substance.

Although the chemical philosophy of the Paracelsians sought to replace the total natural philosophy of the Aristotelians, in fact, most of their efforts were directed toward medical problems. Paracelsus believed that each bodily organ acted as an alchemist separating pure from impure. Thus, the stomach separated the nutritional part of foodstuffs from the dross that was eliminated through the intestines. Similarly, other organs had their function in maintaining the health of the body. Illness occurred when the directive force in an organ failed and poisons accumulated. Examples were the tartaric diseases where stony precipitates developed in the kidneys or bladder, or, as in the case of tuberculosis, in the lungs.

The essentially localized seats of diseases of the Paracelsians differed from the humoral explanations of the Galenists.[20] Utilizing the ancient concept of four humors (blood, phlegm, yellow and black bile) which were associated with the elements, the latter argued that health derived from a proper balance of these fluids while disease was a result of imbalance. The physician might note an excess of blood from a ruddy complexion, yellow bile through the yellowing associated with jaundice, black bile through diarrhoea, or phlegm from a running nose. Even uroscopy might be employed to diagnose an illness by means of a sample of urine without examining the actual patient, since a humoral excess would be evident in the sample. The Paracelsian rejection of humoral medicine was clearly a fundamental break with medical tradition.[21] No less so was their method of cure. The Galenists argued that contraries cure. That is, a disease of a certain quality and magnitude would be cured by a medicine of opposed quality and magnitude. The Paracelsians turned rather to folk tradition, arguing that like cures like: a poison in the body would be cured by a similar poison. And if the Galenists charged that

[19] On the aerial niter see Allen G. Debus, "The Paracelsian Aerial Niter," *Isis*, *55* (1964), 43–61. Zbigniew Szdlo has argued for the origin of this concept in the work of Michael Sendivogius. See, for instance, his "The Alchemy of Michael Sendivogius His Central Nitre Theory," *Ambix*, *40* (1993), 129–146 and "The Influence of the Central Nitre Theory of Michael Sendivogius on the Chemical Philosophy of the Seventeenth Century," *Ambix*, *43* (1996), 80–96.

[20] On Galenic humoral medicine see Owsei Temkin, *Galenism: Rise and Decline of a Medical Philosophy* (Ithaca/London: Cornell University Press, 1973), pp. 17–18.

[21] On disease theory, see Walter Pagel, "Van Helmont's Concept of Disease–To Be or Not To Be? The Influence of Paracelsus," *Bulletin of the History of Medicine*, *46* (1972), 419–54.

they were a legion of homicide physicians, they replied that their medicines were safe because they had been altered chemically and that they paid careful attention to dosage.

These medical reformers not only proposed a new approach to cure; they also emphasized a new class of materia medica.[22] To be sure, some metallic and mineral substances had been employed by the ancients, but the great bulk of traditional remedies were derived from plant substances. This balance was to shift with the chemists who did not eliminate plant-based medicinals, but who argued that the new and violent diseases of their age required stronger medicines. Neither the medieval herbals nor the works of the ancients described substances that could combat syphilis and other new diseases successfully. The internal use of metals and their compounds seemed essential to them. Used as purges and vomitives, their action was truly more violent than the old herbal mixtures. In some cases the new medicines proved too strong and the Galenists accused their opponents of murder. When we examine the chemical and pharmaceutical books of the late sixteenth and the seventeenth centuries, we see directions for the preparation of numerous compounds of mercury, lead, arsenic, and antimony, almost all of which would be avoided today.

THE PARACELSIAN DEBATES

There was then good reason for the confrontation that occurred between the Galenists and the Paracelsians.[23] The latter demanded total educational reform. The natural philosophy of the Aristotelians was to be replaced by a new knowledge of nature founded on fresh observations and a close attention to indisputable Biblical authority. The newly recovered Galenic texts–adored by the medical establishment–were scorned by the Paracelsians, who saw greater truths in the folk medicine of the countryside than in the works of the ancient and Arabic physicians. Furthermore, logic and the use of mathematical reasoning was to be replaced by chemical operations in the laboratory. Henceforth the truths of nature were to be sought through the use of the fire, the furnace, and chemical operations rather than ancient texts and academic disputations.

---

[22]Robert P. Multhauf, "Medical Chemistry and 'The Paracelsians'," *Bulletin of the History of Medicine, 28* (1954), 101–26.

[23]On the Paracelsian Debates see Debus, *Chemical Philosophy, 1*, pp. 127–204.

Still, it was not until the works of Paracelsus began to be published in quantity that much notice was taken of them. The search for his texts began in the 1550s and by the end of the following decade a great number of his works had appeared, resulting in an ever increasing number of disciples. It was soon evident that his books and tracts lacked an overall cohesive plan and this led to a number of works that attempted to supply this need. This was the goal of Peter Severinus, who defended Paracelsian doctrine against the Galenists. In the *Idea medicinae philosophicae* (1571) he argued that Galen had failed because he sought to make medicine a logically-based science in the same way that Aristotle had proceeded with so many other subjects.[24] For him this was a "mathematical" approach, totally unsuitable for medicine. This subject depended on observation, not logic. And here the new chemical medicine of the Paracelsians offered real hope. Severinus attacked Galenic humoral pathology and found truth in the doctrine of signatures and the macrocosm-microcosm analogy. The *Idea* was a much needed defense of the Paracelsian system and it became a standard statement of the new medical system both for its proponents and for its detractors.

Severinus may have been the most important of the early defenders of the Paracelsian system, but there is little doubt that Thomas Erastus (Liebler) (1524–1583) was its most learned opponent.[25] Originally a professor of medicine at Heidelberg, he became more famous after 1580 when he became professor of theology and moral philosophy at Basel. Erastus was willing to accept some of the newer chemically prepared medicines, but little beyond that. He felt that humoral medicine was one of the great achievements of antiquity and that the four Aristotelian elements were indeed the fundamental substances of our world. To have introduced the three principles was a grievous error. Nor was there any truth in the macrocosm-microcosm analogy. To have rejected the belief that contraries cure had led to the introduction of lethal metallic poisons in the mistaken belief that they would cure the sick. Rather, they have led to coffins. Perhaps most dangerous for the

[24]Severinus, *Idea*, pp. 2, 3, 21.

[25]See Debus, *Chemical Philosophy*, *1*, pp. 131–34. An important recent discussion of Erastus will be found in Charles D. Gunnoe, "Erastus and Paracelsianism: Theological Motifs in Thomas Erastus' Rejection of Paracelsian Natural Philosophy" in Allen G. Debus and Michael T. Walton, *Reading the Book of Nature: The Other Side of the Scientific Revolution* (Kirksville, Missouri: Thomas Jefferson University Press, 1998), pp. 45–66. Jole Shackelford has discussed the "Early Reception of Paracelsian Theory: Severinus and Erastus" in the *Sixteenth Century Journal*, *26* (1995), 123–135.

Christian was the link the chemists made between their natural philosophy and biblical truth. The divine Creation was not a chemical separation and the world should not be interpreted chemically. Erastus's *Disputationes de medicina nova Paracelsi* appeared in four parts between 1572 and 1574 and immediately became a standard text for Galenists.

There were, as might be expected, some authors who sought a middle ground. One of these was Johannes Guinter of Andernach,[26] the medical humanist who had taught Vesalius and Servetus at Paris in the 1530s. A seemingly tireless scholar, he began to read the Paracelsian texts as they were published. The result was a large work on the old and new medicines that was published three years before he died. Here one found no acid attack on the ancients. Rather, Guinter maintained their authority in regard to theory. And he had mixed thoughts regarding Paracelsus and his followers. They were often arrogant men who mixed truths and falsehoods together. But many of the new medicines had accomplished marvelous cures that were almost divine. For this reason it was necessary for physicians to consider them carefully. Guinter sought to minimize the differences between the two systems. The Paracelsians denied that "contraries cure," but surely the chemical operations that chemists used to alter substances might change their "similar" to "contrary" properties. And if the Galenists objected to the new emphasis on macrocosm-microcosm analogies, they should recall that this two-world analogy was to be found in the ancient texts as well. In short, he believed that the "ancients on account of time-honored authority are to be given first place," but that there was much of value in the work of the more recent chemists. Physicians should take from both whatever they found of value. It is little wonder that later Paracelsians referred to this work by one of the most prominent Galenists of the century as a testimonial for their own views.

Many physicians of the period were to reflect Guinter's moderate position. Johannes Albertus Wimpenaeus (fl. ca. 1570), Daniel Sennert (1572–1637), and John Cotta (ca. 1575–ca. 1650) all had reservations about the wholesale rejection of ancient medical theory. Cotta (1612) was distressed that "the innumerable dissentions amongst the learned concerning the Arabicke and Chymicke remedies at this day infinitely, with opposite and contradictorie writings, and inuectives, burthen the whole world." Galen

[26] I have discussed Guinter's *De medicina veteri et noua tum cognoscenda, tum faciunda commentarij duo* (1571) in *The Chemical Philosophy, 1*, pp. 139–45.

himself had chosen the best from all authors and his example should be followed.[27]

If we pause to examine a few of the most influential medical authors who touched on this medical "heresy" in the early seventeenth century, we find a broad spectrum of reaction. The "Admonitory Preface" to Oswald Croll's (ca. 1560–1609) *Basilica Chymica* (1609) presents the reader with a theoretical defense of the new medicine.[28] The author, chief physician to Prince Christian I of Anhalt-Bernberg, discussed the relation of Paracelsian medicine to the macrocosm-microcosm universe. The remainder of the volume presented the new medicines with a strong emphasis on the preparations of antimony. Appealing both to the practical pharmacist and the medical theoretician, the *Basilica Chymica* went through numerous translations in many editions into the eighteenth century.

Also interested in pharmaceutical chemistry, Andreas Libavius (1540–1616) was at the same time a classicist who served as the town physician in Rothemburg. Convinced of the value of transmutational alchemy as well as the practical value of medicinal chemistry, Libavius presented typical allegorial plates in his folio volumes. But if he accepted the truths of alchemical imagery, he opposed the rising tide of the mystical philosophies. To this end

[27]John Cotta, *A Short Discoverie of the Unobserved Dangers of severall sorts of ignorant and unconsiderate Practisers of Physicke in England* [London: (R. Field) for W. Jones and R. Boyle, 1612], pp. 82–83. Although he did not comment on the work of Paracelsus, one might also include Jean Fernel (1485–1558) to this list. Known primarily as a Galenist, Fernel was frequently cited with approval by Paracelsians for his views on astrology and the occult causes of disease. In his *De abditis rerum causis* (1548), Book 2, Chapter 8 ("Plurimis functiones & opera in nobis, occultis ex causis esse") he discussed subjects such as the pulse and the formation of the foetus. Jean Fernel, *Vniversa Medicina, cum notis, observationibus, et remedijs secretis* [Traiecti ad Rhenum (Utrecht): Typis Gisbert à Zijll, & Theodori ab Ackersdyck, 1656), *2*, pp. 480–485. This edition has notes and observations by John and Otto Heurn. The subject is extended to disease in following chapters and even takes up sympathy and antipathy in chapter 17 (p. 526). The standard work on Fernel is that of C. S. Sherrington, *The Endeavour of Jean Fernel* (Cambridge, 1946) to which must be added the recent work of Laurence Brockliss and Colin Jones, *The Medical World of Early Modern France* (Oxford: Clarendon Press, 1997), pp. 128–138 and *passim.*

[28]Oswald Crollius, *Basilica Chymica* (Frankfurt: Apud Claudium Marnium & heredes Joannis Aubrii, 1609). Owen Hannaway compared the work of Crollius and Libavius in *The Chemists and the Word: The Didactic Origins of Chemistry* (Baltimore: The John Hopkins Press, 1975). Hannaway then went on to discuss "Laboratory Design and the Aim of Science: Andreas Libavius versus Tycho Brahe," *Isis*, 77 (1986), 585–610. Here he pointed to Tycho's practice of chemistry as a guide to cosmic harmony in contrast to the use of the laboratory by Libavius for the products of the furnaces and their benefit for mankind. Jole Shackelford replied to Hannaway's article in "Tycho Brahe, Laboratory Design and the Aim of Science: Andreas Libavius versus Tycho Brahe," *Isis*, *84* (1993), 211–230 in which he rejected the sharp contrast seen by Hannaway and argued that both the approach of Tycho and Libavius contributed to modern science.

he rejected the work of Paracelsus and later the neo-Paracelsian Rosicrucian texts. Libavius may have been a believer in Aristotelian philosophy, but he strongly supported the use of chemicals in medicine.

Libavius was opposed in England by the hermetic philosopher Robert Fludd (1574–1637), who sought to replace the traditional educational system with a new one based on Biblical authority and mystical truths.[29] He countered Libavius's attack on the *Fama Fraternitatis* and the *Confessio* of the Rosicrucians (1615) with a call for reform of higher education that would take the true student beyond the fundamentals of given subjects to inner truths known by ancient philosophers. To this he added a belief in the macrocosm-microcosm universe. To some early seventeenth-century scholars, his folio volumes on the great and small worlds seemed to offer the true answer to the ancients. But if Fludd routinely used chemical examples, he referred to Paracelsus only occasionally. He was, indeed, a chemical philosopher, but surely more of a hermeticist than a Paracelsian.

Fludd's work truly united religion with his views on cosmology and medicine. This connection remained a problem for others. The professor of medicine at Wittenberg, Daniel Sennert (1572–1637), discussed the current debate in 1619 in his *De chymicorum cum Aristotelicis et Galenicis consensu ac dissensu.* He strongly supported the traditional humoral medicine while at the same time calling for the adoption of the Paracelsian principles. For him the new chemical medicines were an important addition to the materia medica even though chemistry itself could not be allowed to dominate medicine. Above all he criticized those who sought to combine chemistry and religion. Those who do this present

> a "false chemistry" which hath its particular Religion, for because they think they have reformed or perfected all Philosophy & Physick, they stay not there, but proceed to Divinity, and mix prophane and holy things together, and so, they bring in any absurdity, and dispute wonderfully of the Kingdom of the Blessed, the Angels Miracles, Faith and Resurrection.[30]

---

[29]In addition to my discussion of Fludd in *The Chemical Philosophy, 1,* pp. 205–93, see William H. Huffman, *Robert Fludd and the End of the Renaissance* (London and New York: Routledge, 1988); Joscelyn Godwin, *Robert Fludd: Hermetic Philosopher and Surveyor of Two Worlds* (London: Thames and Hudson, 1979).

[30]Daniel Sennert, Nicholas Culpeper, and Abdiah Cole, *Chymistry Made Easie and Useful. Or, The Agreement and Disagreement of the Chymists and Galenists* (London: Peter Cole, 1662), p. 24. For the Latin see Daniel Sennert, *De chymicorum cum Aristotelicis et Galenicis consensu ac dissensu liber 1. Controversias plurimas tam philosophis quam medicis cognitu utiles continens* (Third edition, Paris: Apud Societatem, 1633), p. 57. The first edition was published at Wittenberg in 1619.

The views we see expressed by Sennert, Fludd, Croll, and others indicate the breadth of interest in chemistry early in the seventeenth century. As this interest spread throughout Europe, it met with varying degrees of opposition from the local Galenic establishment. This reaction may have been most severe in France, where the Parisian medical faculty attempted to maintain Galenic authority in medicine.[31] Paracelsus was first noted in the French medical literature in a translation of Mattioli's *Commentaries* on Dioscorides (1561). Here Paracelsus was associated with the internal use of antimony as a purgative. This was noted with approval by an obscure physician of La Rochelle, Loys de Launay (fl. 1557–1566), who wrote a short tract praising the remedy (1564). Jacques Grévin (1540–1570), a member of the Medical Faculty, replied to him two years later, charging that in reality antimony in any form was a dangerous poison. Presenting his case to the assembled members of the Medical Faculty, he won from them a decree that the substance was a poison that should not be taken internally (3 August 1566).

This early exchange was the beginning of the aptly termed "Antimony War" that was to continue for a full century and was to affect the relationship of chemistry and medicine in every part of Europe from Scandinavia in the north to Italy in the south, and from Portugal in the west to the Ottoman Empire in the east. In France the chemists never flagged in their attacks on medical authority, while the Medical Faculty never lost an opportunity to eliminate the challenge to tradition. In 1578 they expelled the Norman Paracelsian, Roch le Baillif (fl. 1578–1580), from Paris in an attempt to maintain their authority. Unfortunately for the learned doctors, no amount of legal action seemed to prevent the ever-increasing interest in this subject. Chemical books continued to be published praising the new medicines and condemning the medical views of the ancients.

Perhaps the most important of the late sixteenth-century French Paracelsians was Joseph Duchesne (Quercetanus) (ca. 1544–1609), who began to publish his views on the importance of chemistry as early as 1575. After Henry of Navarre took Paris and was crowned as Henry IV (1593), Duchesne was appointed a royal physician by Jean Ribit (ca. 1571–1605), a Paracelsian who was the chief physician to the King. Ribit was later to appoint Theodore Turquet de Mayerne (1573–1655) as a royal physician. Mayerne was to play

[31]See Allen G. Debus, *The French Paracelsians: The Chemical Challenge to Medical and Scientific Tradition in Early Modern France* (Cambridge University Press, 1991). The debate over the use of antimony is described in Pascal Pilpoul, *La Querelle de l'Antimoine (Essai historique).* Thèse pour l'Doctorat en Médecine (Paris: Librairie Louis Arnette, 1928).

an important role in the development of English chemical medicine. All three of these French iatrochemists were Protestants, a matter of additional concern to the Parisian medical establishment, since the Sorbonne required religious orthodoxy from medical students.

Both Duchesne and Mayerne sought to establish pharmaceutical chemistry through courses that taught the proper preparation of chemicals. This move was also opposed by the Medical Faculty. However, the conflict reached a new level of bitterness with the publication of Duchesne's *Liber de priscorum philosophorum verae medicinae materia* in 1603. Here he offered a comprehensive chemical interpretation of man and the cosmos, one that was seen by the Parisian doctors as a frontal attack on the official Galenism. Duchesne was answered first by Jean Riolan the Elder (1539–1606) and then defended by his friend, Turquet de Mayerne. The latter's *Apologia* (1603) so offended the members of the Faculty that he was ostracized by them. Those who ignored this decree were threatened with the loss of their own medical degrees and the privileges of the Faculty. It is little wonder that Mayerne was eventually to leave France for England.

Although the Medical Faculty forbade the sale of all chemical remedies in France in 1615, this decree was impossible to enforce. The value of chemical preparations continued to be discussed in a stream of publications, and the approval of a Royal Botanical Garden included the provision for a professor of chemistry who would demonstrate the preparation of medicinal chemicals. Although the Garden opened in 1640, it was another eight years before the noted Scot, William Davisson (ca. 1593–1669), took up his post as the first professor of chemistry. He was only the first in a line of distinguished chemists at the Jardin des Plantes during the course of the seventeenth and eighteenth centuries.[32]

Somewhat earlier Théophraste Renaudot (1584–1653) had established his Bureau d'Adresse in Paris (ca. 1630) where he dispensed free medical advice, held weekly conferences that included scientific and medical subjects, and recommended the use of chemical medicines. Perceived as a threat by the Medical Faculty, the members instituted legal proceedings against him when he proposed to expand his facilities. With the death of his patron, Car-

[32]Still a fundamental source for French chemistry in the seventeenth century is Hélène Metzger, *Les doctrines chimiques en France du début XVIIe siécle. Premiere partie* (Paris: P. U. F., 1923). On chemistry at the Royal Botanic Gardens see Rio Howard, "Guy de la Brosse and the Jardin des Plantes in Paris" in *The Analytic Spirit: Essays in the History of Science in Honor of Henry Guerlac*, ed. Harry Woolf (Ithaca: Cornell University Press, 1981), pp. 195–224.

dinal Richelieu, in 1642, Renaudot had lost his main supporter. He lost his case and was stripped of nearly all of his privileges.[33]

With the defeat of Renaudot, the Medical Faculty had seemingly triumphed in their long battle to prevent the use of chemistry in medicine. However, the older Galenists were aging and antimony was becoming an increasingly fashionable remedy. After Louis XIV was cured with an antimony purge (1658), the end was in sight. An assembly of the Medical Faculty in 1666 resulted in an overwhelming vote (92 to 10) favoring the addition of emetic wine to the list of approved purgatives, and after that time there was little resistance to the use of chemical medicines along with more traditional cures.

In England as in France, news of the new chemical medicine began to spread in the 1560s due in part to the fact that some of the Marian exiles had settled in the Swiss Cantons.[34] Interest in the anti-Galenic remedies spread rapidly. Conrad Gesner's book on distillation appeared in English translation and dedicated surgeons and distillers published works promoting the virtues of remedies prepared in the new fashion.

The first defense of Paracelsian theory in England was that of Rychard Bostok, who compared the work of the chemists with that of the Galenists in 1585.[35] For Bostok the work of Paracelsus, Copernicus, Calvin, and Luther signalled a return to God's truth in religion, natural philosophy, and medicine. He referred his readers to a long list of Continental authorities who followed Paracelsus. A second early English convert was Thomas Moffett (1553–1604), who had taken his M.D. at Basel in 1578.[36] A few years later he was in Denmark, where he became acquainted with both Peter Severinus and Tycho Brahe (1546–1601). Moffett's work on chemical medicine appeared in 1584 and was later to become a standard text through its inclusion in Lazarus Zetzner's *Theatrum Chemicum* (1602). Dedicated to Severinus, Moffett's volume defended the use of chemically-prepared medicines, the interconnection of man and the great world, and the three Paracelsian principles. It was a learned defense of the Paracelsian system and surely the reason why Moffett

---

[33]On the Renaudot affair and the eventual triumph of antimony see Debus, *French Paracelsians*, pp. 84–99.

[34]Allen G. Debus, *The English Paracelsians* (London: Oldbourne, 1965).

[35]*Ibid.*, pp. 57–65. On Bostok see David Harley, "Rychard Bostok of Tandridge, Surrey (c. 1530–1605), M.P., Paracelsian Propagandist and Friend of John Dee," note 10 above.

[36]Debus, *English Paracelsians*, pp. 71–76. Moffett's connection with Severinus is discussed in the essential work on Danish Paracelsism by Jole R. Shackelford, *Paracelsianism in Denmark and Norway in the 16th and Seventeenth Centuries* (University of Wisconsin-Madison Doctoral Dissertation, 1989).

was appointed to a committee of the College of Physicians then considering a section on chemical medicines for a planned pharmacopoeia of the College.

A third defense of chemical medicine was printed in England in 1605 when Thomas Tymme translated long extracts from Joseph Duchesne's works on the new medicine that had been published in 1603 and 1604.[37] The influence of the French debate intensified when Theodore Turquet de Mayerne left Paris in 1606 for England, where he was incorporated M.D. at Oxford.[38] After a temporary return to France, he moved to England permanently after the death of Henry IV in 1610. He was then appointed First Physician to James I and was elected a Fellow of the Royal College of Physicians in 1616. Here he involved himself in the long dormant pharmacopoeia project, pressing for the inclusion of chemically-prepared medicines. The preface to the *Pharmacopoeia Londinensis* of 1618–surely written by Mayerne–states that

> we venerate the age old learning of the ancients and for this reason we have placed their remedies at the beginning, but on the other hand, we neither reject nor spurn the subsidiary medicines of the more recent chemists and we have conceded to them a place and corner in the rear so that they might be as a servant to the dogmatic medicine, and thus they might act as auxiliaries.[39]

It is clear that the bitter debate in France over the acceptance of the new medicines resulted in the departure of Turquet de Mayerne for London, where he became influential in including these chemicals in the first national pharmacopoeia. However, it should be noted that Mayerne also played a part in establishing the primacy of the College over other medical groups. The publication of the *Pharmacopoeia Londinensis* coincided with the Royal Charter to the College (1618). This restated the right of the College to search and examine all apothecary wares and extended this right to include the "Distillers and Sellers of waters or oyles" as well as the "Preparers of Chymi-

---

[37]Debus, *English Paracelsians*, pp. 87–97.

[38]On Turquet de Mayerne and the *Pharmacopoeia Londinensis* see *ibid.*, pp. 150–56. This second issue of the *Pharmacopoeia* supplanted an earlier one published in May, 1618. The passage remains the same in both. A facsimile of the first issue with a comparison of the two is to be found in the *Pharmacopoeia Londinensis of 1618 Reproduced in Facsimile with a Historical Introduction by George Urdang*, Hollister Pharmaceutical Library Number Two (Madison: State Historical Society of Wisconsin, 1944), p. 93

[39]*Pharmacopoeia Londinensis* (2nd ed., London: Iohn Marriott, 1618), "Candido Lectori."

cal Medicynes." Twenty years later Mayerne sought to further regulate the distillers by the founding of a City Company of Distillers.

In short, although interest may have been centered in Germany and Switzerland, chemical medicine became widely known in England, France, and other European countries in the course of the sixteenth and early seventeenth centuries. This influence was even to be found in the Ottoman Empire, where an Arabic work titled the *New Chemical Medicine Invented by Paracelsus* was completed by Salīh Ibn Nasrallāh Ibn Sallūm (d. 1669/1670) no later than 1640. The relatively large number of manuscript copies of this work attest to the fact that the Paracelsian union of chemistry and medicine had spread beyond the borders of Western Europe by the mid-seventeenth century.[40]

CHEMISTRY AND THE UNIVERSITIES

If chemistry spread throughout Europe in the course of the century, it also made inroads within the university system. Here Galenism had been solidly entrenched in the medical faculties and the early Paracelsians felt that their chemical knowledge had been effectively ruled out of any consideration by scholars. Some chemists boasted that the reason why they were learned is that they had avoided the universities entirely. Johann Rudolph Glauber (ca. 1603–1670), revered by chemists no less than van Helmont in mid-century, wrote that

> I confess ingeniously, that I never frequented the Universities, nor ever had a mind to do so; for should I have so done, haply I should never have arrived to that knowledge of Nature, which I mention without boasting, as I now possess; neither doth it ever repent me, that I have put my hands to the Coals, and have by the help of them penetrated into the knowledge of the secrets of nature . . .[41]

[40] Allen G. Debus, "Mysticism and the Rise of Modern Science," *Journal of Central Asia*, *3* (1980), 46–61 (56–58).

[41] Johann Rudolph Glauber, *The Prosperity of Germany* (Part I) in *The Works . . . containing, Great Variety of Choice Secrets in Medicine and Alchymy . . .*, trans. Christopher Packe (London: Thomas Milbourn, 1689), *2*, p. 373.

But if the universities either ignored chemistry or were openly hostile to it, students were still able to find instruction. If the subject was anathema to the professors of the Sorbonne, it was not at the second medical school in France, Montpellier, where a long interest in chemistry derived from the early work of Arnald of Villanova. But even in Paris, students could find private instruction. Jean Beguin (1550–1620) established a laboratory there with the encouragement of Jean Ribit and Turquet de Mayerne. Here he not only gave lectures on pharmaceutical preparations, but also prepared the first true chemical textbook, the *Tyrocinium chymicum* (1610).[42] This was to become the model for later texts with its emphasis on practical processes and the use of chemical equipment. For Beguin the preparation of medicines with the use of fire made them more valuable than the traditional herbal mixtures of the Galenists because the heat employed had separated pure essences from their useless impurities.

Beguin's text enjoyed enormous popularity, with more than forty editions printed to 1690. It was embellished with notes and commentaries by a number of authors and translated into many European languages. But soon others began to prepare their own texts.[43] The founder of the Jardin des Plantes, Guy de la Brosse (ca. 1586–1641), included a lengthy discussion of chemistry in his *De la Natur, Vertu, et Vtilité des Plantes* (1628), and after the appointment of William Davisson (ca. 1593–ca. 1669) as the first Professor of Chemistry at the Jardin, we find a succession of chemical textbooks written for the various lecture series presented. Davisson's *Philosophia Pyrotechnia seu Curriculus Chymiatricus* appeared in two parts (1633, 1635) and later was translated into French. His successors at the Jardin des Plantes, Nicholas Le Fèvre (ca. 1615–1669), Christofle Glaser (ca. 1615–ca. 1672), and Moyse Charas (1619–1698) continued in this lecturer-author tradition throughout the rest of the century. The tradition culminated in the *Cours de chymie* of Nicholas Lemery (1645–1715), which appeared in French in numerous editions from 1675 to 1757 and was translated into Latin, German, English, and Spanish.

The texts of these authors as well as others written by private instructors who had no connection with the Jardin followed a general pattern even though they differed on some theoretical points. They used similar arrangements of

---

[42]T. S. Patterson, "Jean Beguin and his *Tyrocinium Chymicum*," *Annals of Science*, *2*, (1937), 243–98. For a discussion of the relation of the Paracelsian tradition in France to medical institutions see Laurence Brockliss and Colin Jones, *The Medical World of Early Modern France* (see note 27 above). Here the authors point out that iatrochemistry was discussed in faculty theses by the mid-1660s.

[43]On the French textbooks see Metzger, *op. cit.* and Debus, *The French Paracelsians*, pp. 123–24.

their material beginning with general information and then going on to a description of chemical equipment and processes prior to the actual division into mineral vegetables and animal substances. Special emphasis was always placed on the medicinal properties of the preparations and because of the interest in antimony as a purge, its chemistry was generally discussed extensively.

The large number of chemical textbooks and their many editions in the course of the seventeenth century surely indicate the intense interest in this subject. Given this interest it would have been surprising if chemistry had not made inroads at the universities. However, this did not happen as the sixteenth-century Paracelsians might have wished. Although emphasizing medicine, the Chemical Philosophy was proposed first as a replacement not only for ancient medicine, but for the total natural philosophy of the schools. But interest in the macrocosm-microcosm analogy declined while the medical implications continued to grow. As a result we find chemistry being established in the European medical faculties in the course of the seventeenth century, while the natural philosophy curriculum remained wedded primarily to subjects we would classify as the physical sciences.

Although some specific chemical preparations for pharmaceutical purposes were surely taught in many places in the course of the sixteenth century, the first chair created specifically for that purpose came from the Medical Faculty at the University of Marburg in 1609.[44] The first professor of Chymiatria, Johann Hartmann (1568–1631) was a Paracelsian, but his teaching was largely directed to pharmaceutical preparations. Hartmann also prepared editions of the works of Oswald Croll and Jean Beguin, which were popular throughout Europe. Hartmann's medical appointment was the first of many to come. At Jena, Zacharias Brendel (1553–1626) offered a course in chemistry to medical students as early as 1612. His son, also named Zacharias (1592–1638), continued this course and in 1641 Werner Rolfinck (1599–1673) became the first Professor of Chemistry in the Medical Faculty. He held this position for decades and because of this long tradition, Jena became a major school for training in chemistry. At Jena as elsewhere the professors prepared their own textbooks, which generally followed the tradition already established in France. During the second half of the century, the medical fac-

[44]I have discussed the introduction of chemistry to the European universities in "Chemistry and the Universities in the Seventeenth Century," *Academiae Analecta: Klasse der Wetenschappen, 48* (1986), 13–33. Bruce T. Moran is examining the work of Johann Hartmann in depth in a series of books and papers. Particularly useful is his *Chemical Pharmacy Enters the University: Johannes Hartmann and the Didactic Care of Chymiatria in the Early Seventeenth Century* (Madison: American Institute of the History of Pharmacy, 1991).

ulties of many Central European universities founded chairs in chemistry: Wittenberg, Helmstedt, Leipzig, and Halle among them.

In the Netherlands the preeminent medical school in the sixteenth and the seventeenth centuries was at Leiden, but it was not until 1669 that Carel de Maets (1640–1690) was appointed as an unsalaried docent to teach chemical preparations to medical students. Both he and his successor had worked in Glauber's laboratory in Amsterdam. At Leiden the early instructors prepared chemical textbooks that show that the chemistry being taught there was centered on pharmaceutical compounds.

At Oxford and Cambridge chemical instruction began in 1683, but again it is the French scene that is particularly illuminating. As we have seen, the strong opposition of the Parisian medical faculty had resulted in the development of chemical courses outside of the academic curriculum. The lectures at the Jardin des Plantes led the way in the teaching of chemistry not only in France, but in Europe. Those interested in the subject could attend the lectures there or go to private instructors. It was not until the end of the century in 1696 that the Medical Faculty in Paris established a fifth professorial chair for the teaching of both chemical and Galenic pharmacy–a date considerably later than other medical schools. Even Montpellier, often identified by the Parisian doctors as a seat of chemical heresy, did not appoint a demonstrator in chemistry until 1673.

In short, chemistry was well established in European universities by mid-century and it had become almost universal by the end of the century. However, it had been accepted as a branch of medicine rather than as the basis of natural philosophy. Its medical orientation was to affect its immediate future.

THE CHEMICAL DREAM AND THE NEW PHILOSOPHY

Although chemistry clearly had its greatest impact on medicine in the sixteenth and the seventeenth centuries, it should not be forgotten that for Paracelsus and many of his followers the chemical philosophy was to be the basis of a new understanding of both man and nature. It was for them *the* new philosophy that would supplant the ossified works of Aristotle, Galen, and all the rest of the ancient authors.

The reaction against this chemical philosophy may be seen best in the astonishing (to us) attention paid to the work of Robert Fludd by Johannes

[45]Debus, *Chemical Philosophy, 1*, pp. 256–279.

Kepler (1571–1630), Marin Mersenne (1588–1648), and Pierre Gassendi (1592–1655).[45] We have already referred briefly to Fludd in his attack on Andreas Libavius and his support of the elusive Rosicrucian Brotherhood (1616–1617). It is clear that he sought to establish an all-embracing hermetic philosophy that would be closely connected with religion. He was a physician and would have considered himself a chemical philosopher, although he rarely referred to Paracelsus.

In contrast to many of the chemical physicians of his day, Fludd had an impeccable university training. He was born to a well-to-do family and educated at Oxford University. After taking his B.A. (1597) and his M.A. (1598), Fludd spent six years on the Continent before returning to England, where he was incorporated M.D. at Oxford. He then settled in London, where he became a prominent physician at Court. Although he repeatedly offended the Fellows of the College of Physicians of London with his outspoken views on the chemical cosmology, he was admitted to that College in 1609.

His publications against Libavius were only the beginning of a torrent of works that appeared throughout the remainder of his life. In these works and in the debates they engendered, we may find key issues being discussed that affected the development of the new science of the seventeenth century.

Although Fludd's replies to Libavius were an apology for the Rosicrucians, they were also an attack on the state of higher education with its emphasis on Aristotle and Galen. Fludd much preferred to rely on the Holy Scriptures, the *Corpus Hermeticum*, the more mystical Platonic and Pythagorean authors, and chemical observations and analogies. He sought an understanding of the world through the macrocosm and the microcosm and he thought that the unifying principle was the Divine Spirit which gave life to all things. He proposed a plan that was essentially a research program that would investigate the action of this spirit in the macrocosm and the microcosm. Indeed, much of Fludd's subsequent work was directed to this study. In his *Philosophicall Key* (c. 1618) he sought to isolate this spirit from wheat, and in his series of works on the macrocosm and the microcosm he discussed the Creation and the divine harmonies as they affected man and nature. He described the Creation as a chemical process of separation and repeatedly presented observations and experiments that would support his view.

His description of the harmonies that ruled the heavens (1617) appeared in print just as Johannes Kepler was completing his *Harmonices mundi* (1619). Fludd's belief in a universal monochord that indicated the relative distances between the sun, the moon, and the planets was far removed from the relationships presented in Kepler's *Harmonices*, to which he had added a refutation of Fludd's views on this subject. Although Kepler believed in the significance of music as a key to the heavens, his attitude toward mathematics was quite different from that of Fludd. Kepler felt that any general theory must be upheld both by his observations and by mathematical calculations, while Fludd was so convinced of the truth of his concepts that he considered anything else of secondary importance. The debate between the two continued until 1622 and contributed to an understanding of the role of mathematics in the sciences.

Fludd's chemical interpretation of the Creation and his use of chemical analogies throughout his discussion of the macrocosm and man led to a second debate with Marin Mersenne and Pierre Gassendi. Both were admirers of Galileo and the use of mathematical abstraction in the sciences and both had been concerned about the current interest in France in alchemy and in the Rosicrucian texts. In 1625 Mersenne published his *La Verité des Sciences* in dialogue form. Here an alchemist argued that chemistry should be the fundamental science since it is based on fresh observations. Needless to say, it was Mersenne as the "Christian Philosopher" who won the argument, suggesting that many of the views of the chemists were either doubtful or quite wrong. The chemists, he said, should clarify their language and stay clear of allegories and chemical interpretations of Scripture. Indeed, their claims to be the best interpreters of nature should be rejected. Moreover, chemistry itself should be policed by a chemical academy that would encourage projects that dealt primarily with the development of new medicinal substances.

Mersenne had attacked Fludd in print earlier and the latter's reply initiated another debate. Fludd wrote that it was astonishing that Mersenne as a man of God could not understand that the work of the chemist was directed to a better understanding of the work of the Creator. He faulted Mersenne for confusing the works of alchemical charlatans with those of the true chemists whose goals were noble and closely associated with those of religion. In fact, the two did not understand each other. Mersenne then sent a collection of Fludd's works to Pierre Gassendi with the hope that his friend could write a convincing reply to them. Gassendi's examination of Fludd's philosophy was published in 1630 and included a rejection of William Harvey's (1578–1657) work on the circulation of the blood (1628), as both Mersenne and Gassendi seem to have thought that Harvey was a disciple of Fludd. In his reply to Gassendi (1633), Fludd argued correctly that Gassendi had rejected the views of Harvey on the basis of a single dissection he had witnessed, while his own views and those of Harvey had been based on many dissections of hearts performed at the Royal College of Physicians.

The importance of Fludd thus rests partially on his own search for a new philosophy grounded on mystical and religious truths. But he is no less important for his willingness to enter into debates with others. The significance of the debate with Kepler centers on the role of mathematics in scientific method, while his debate with Mersenne and Gassendi takes up the role of experiment in the sciences and illustrates the issues separating the early mechanists from a convinced hermeticist and chemical philosopher.

CONCLUSION

The Fluddean debates were neither the first nor the last to deal with the role of chemistry in the interpretation of nature as a whole. However, in touching on the relationship of observation and experiment, the use of mathematics, and the state of the universities in the teaching of natural philosophy as well as medicine, they point to key areas that separated the chemical philosophers from the mechanical philosophers in the seventeenth century. Here were two "modern" schools–both in opposition to the ancients.

And yet, although both chemists and mechanists claimed to hold the key to a new science, it is clear that the chemists were closely linked to medicine while the mechanists were not at this time. The lines were clearly drawn between Galenists and Paracelsians from the start. A prime area of contention was that of the reform of medical education. But beyond that the chemists questioned the traditional elements, sought a principle of cure based on similitude rather than contrariety, and demanded the introduction of an armory of metallic-based remedies. Their rejection of humoral explanations was anathema to the medical establishment, and here again, the turn to fundamentalist and mystical religious interpretations strongly mixed with hermeticism set them apart from other physicians.

In short, by the mid-seventeenth century we can view nearly a century of debate. Many of the questions raised during that period continued to be debated into the eighteenth century, but because chemistry became academically respectable through the medical faculties of European universities, it is in medicine that we find the most interesting developments in the late seventeenth century. It is a period that saw continued debate . . . now primarily between chemical and mechanical physicians (or more properly between

CHAPTER 2

Jean Baptiste van Helmont and the New Chemical Medicine

Let it be a shame, I say, that the Schooles of Natural Phylosophy have more bestowed their Contemplation . . . [on] . . . Science Mathematical or Learning by Demonstration, than on Nature it self . . .[1]

Therefore . . . a fourfold darknesse of Ignorance . . . entred . . . into the profession of Medicine, and they have left it without honour; To wit, the ignorance of Causes, manner of making, of the Remedy, and suitable application thereof. Truly, as the art of the fire, unlocks Bodyes before our eyes; so it opens the Gate unto Natural Philosophy. The true medicine therefore hath layen hid, as depressed under the Ignorance and sluggishnesse of the <u>*Schooles*</u>*: and that preparation of Medicine which ought to bring light unto a Physitian, is wholly accounted mechanical and conferred on the Apothecary and his Wife.*[2]

Jean Baptist van Helmont as translated by John Chandler, 1662

---

[1]John Baptista van Helmont, *Oriatrike or Physick Refined. The Common Errors therein Refuted, And the whole Art Reformed & Rectified. Being A New Rise and Progress of Phylosophy and Medicine, for the Destruction of Diseases and Prolongation of Life*, trans. J(ohn) C(handler) (London: Lodowick Loyd, 1662), p. 636 from "Of Time." For the original Latin, see *Ortus medicinae, Id est, initia physicae inaudita. Progressus medicinae novus, in morborum ultionem, ad vitam longam* (Amsterdam: Ludovicus Elsevir, 1648; reprinted Brussels: Culture et Civilisation, 1966), p. 633. The *Ortus* was issued in 1648 with a companion volume, the second edition of the *Opuscula medica inaudita*, which included the "De Lithiasi," the "De febribus," the "De humoribus Galeni," and the "De peste" (Amsterdam: Ludovicus Elsevir, 1648; reprinted Brussels: Culture et Civilisation, 1966). These three editions will henceforth be cited as *Oriatrike*, *Ortus*, and *Opuscula*.

[2]"De Lithiasi" in *Opuscula*, p. 40; *Oriatrike*, p. 859.

Although van Helmont (1579/80–1644)[3] was a contemporary of Robert Fludd (1574–1637), and although each one viewed his own work as the proper replacement for the ancient authorities, the relative influences of the two differed. Fludd's hermetic, mystical and religious approach to nature found few adherents after his death. On the other hand, van Helmont's emphasis on a more observational science and medicine proved to be more in line with newer scientific methodologies. His massive works were frequently reprinted into the early eighteenth century and they were to spur a new interest in a chemical medicine that was centered less on the preparation of chemicals of pharmaceutical value and more on an understanding of bodily processes.

VAN HELMONT'S SEARCH FOR TRUTH

With van Helmont we have one of the relatively few scientific autobiographies of the seventeenth century.[4] In it he tells of his search for truth through knowledge. The youngest in his family, he was "brought up in Studies" and finished his course in philosophy at Louvain in 1594. Then, reflecting on his training, he was dismayed to realize that he knew nothing.

> I found for certainty, that I was blown up with the Letter, and (as it were the forbidden Apple being eaten) to be plainly naked, save, that I had learned artificially to wrangle. Then first I came to know within my self, that I knew nothing, and that I knew that which was of no worth.[5]

He had studied logic, astronomy, and natural philosophy. The *Elements of Geometry* of Euclid had led him to the Circle of Cornelis Gemma and since this

[3]The most recent full-scale study of van Helmont is that of Walter Pagel, *Joan Baptista Van Helmont; reformer of science and medicine* (Cambridge University Press, 1981). See also Allen G. Debus, *The Chemical Philosophy: Paracelsian Science and Medicine in the Sixteenth and Seventeenth Centuries* (2 vols., New York: Science History Publications, 1977), 2, pp. 295–379. Important recent books are Berthold Heinecke, *Wissenschaft und Mystik bei J.B. van Helmont (1579–1644)* (Bern, Berlin, et al.: Peter Lang, 1996) and Paulo Alves Porto, *Van Helmont e o Conceito de Gás: Química e Medicina no sécolo XVII* (São Paulo: Educ, 1995).

[4]"Studia Authoris," *Ortus*, pp. 16–18; *Oriatrike*, pp. 11–14.

[5]*Ibid., Ortus*, p. 16; *Oriatrike*, p. 11.

> onely commended *Nicholas Copernicus*, I left not off, till I had made the same familiar unto me. Whence I learned the vain excentricities, or things not having one and the same Center, another circular motion of the Heavens; and so I presumed, that whatsoever I had gotten concerning the Heavens, with great pains, was not worthy of the time bestowed about it.
>
> Therefore the Study of Astronomy, was of little or no account with me, because it promised little of certainty or truth, but very many vain things. Therefore having finished my Course, when as I knew nothing that was sound, nothing that was true, I refused the Title of Master of Arts; being unwilling that they should declare me Master of the seven Arts, who was not yet a Scholar. Therefore I seeking truth, and knowledge, but not their appearance, withdrew my self from the Schooles.[6]

Clearly, van Helmont's study of the "Science Mathematical" had resulted in disappointment. Attracted next to the lectures of Martinus del Rio on magic, he turned to that subject, only to gather "empty stubbles, and most poor patcheries, void of judgment." From there he went on to Stoicism and the Christian mystics. In a dream he became aware that understanding came only with the Grace of the Lord, but certainty continued to elude him.[7]

His search continued now in the herbals of Mattioli and Dioscorides. There seemed to be hope in medicine that "might have its Theoremes and chief Authores, instructed by an infused knowledge, into whom . . . the spirit of the Lord had inspired the Causes and knowledge of all Diseases, and also the knowledge of the properties of things."[8] Van Helmont then asked whether there might be some author "who described the Endowments, Properties, Applications, and proportions of Vegetables, from the *Hyssop*, even to the Cedar of *Libanus*?" But no, a professor of medicine told him that what he sought was not to be found in Galen or in Avicenna.[9]

He ruled out the study of law and turned to an exhaustive study of medicine:

---

[6] *Ibid.*, *Ortus*, p. 16; *Oriatrike*, p. 12.

[7] *Ibid.*, *Ortus*, p. 17; *Oriatrike*, pp. 12–13.

[8] *Ibid.*, *Ortus*, p. 18; *Oriatrike*, p. 13.

[9] *Ibid.*, *Ortus*, p. 18; *Oriatrike*, p. 13.

> . . . I read the works of *Galen* twice, once *Hippocrates* (whose Aphorismes I almost learned by heart) and all *Avicen*, and as well the *Greeks*, *Arabians*, as *Moderns* . . . [but they all sang] the same Song, did promise nothing of soundness, nothing that might promise the knowledge of truth, or the truth of knowledge.[10]

Turning to practice, van Helmont accompanied a physician on his rounds, but saw dramatically the uncertain nature of the profession. Then, in another dream, he became convinced that he was to be a physician.[11] And indeed, in 1599 he took his degree in medicine at Louvain. He continued his studies and ten years later moved to Vilvord, where he proceeded "to view the Kingdoms of Vegetables, Animals and Minerals."[12] It was then that he found the works of Paracelsus. He wrote "I admired that man, and too much honoured him: till at length, understanding was given, of his Works, and Errours."[13] There is little doubt that van Helmont had great respect for Paracelsus even though he criticized much of his work.

> For when he (the Lessons of the Ancients being rejected) had sufficiently understood, that there was nothing of Foundation or Truth in Complexions and Humours, he began by variously doubting, to inquire into the most immediate Cause of Diseases, and Posterity owes him Praise for it.[14]

But such admiration, even when tempered with criticism, augured ill for the future. He knew that "by Haters, I am called a Paracelsian, and a Forsaker of the Schools . . ."[15]

Van Helmont's Paracelsism is most evident in his first work, the *De magnetica vulnerum curatione*. A sympathetic cure, often ascribed to Paracelsus, the weapon-salve became a center of debate in the seventeenth century. Briefly, the proponents of this cure asserted that anointing a weapon that had caused a wound with a salve containing the blood of the injured person would lead to healing. Widely accepted as valid, Rudolph Goclenius wrote that the cure

---

[10] *Ibid.*, *Ortus*, p. 18; *Oriatrike*, p. 13.

[11] *Ibid.*, *Ortus*, p. 19; *Oriatrike*, p. 14.

[12] "Promissa Authoris" (Column 3, sec 7), *Ortus*, p. 12; *Oriatrike*, p. 7.

[13] *Ibid.*

[14] "Tria Prima Chymicorum Principia . . . ," *Ortus*, p. 399; *Oriatrike*, p. 402.

[15] "Praefatio," *Ortus*, p. 630; *Oriatrike*, p. 632.

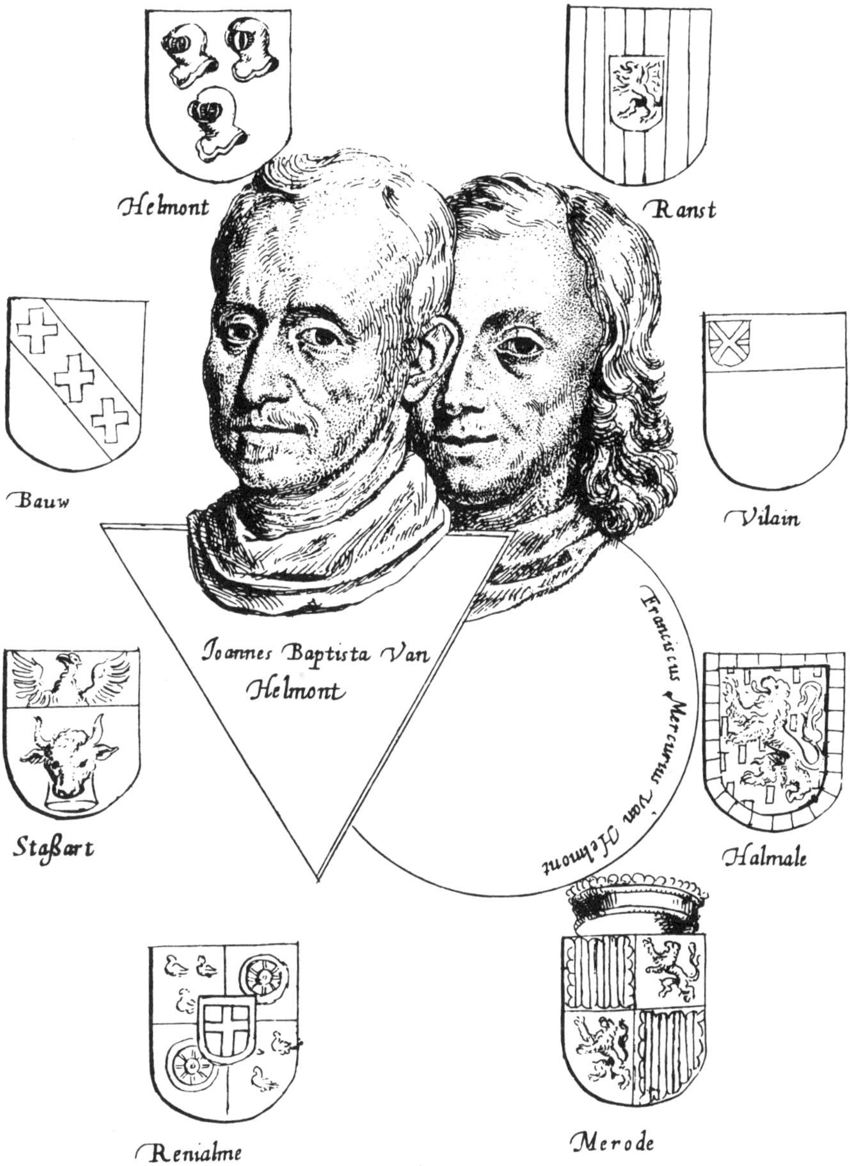

Jean Baptiste van Helmont (1579/1580–1644) and his son, Franciscus Mercurius van Helmont (1614–1699). Frontispiece to the *Ortus medicinae* (Amsterdam: Elsevier, 1648).

occurred through natural causes (1608 and 1613). He was opposed by the Jesuit Jean Roberti, and a series of tracts pro and con were published on this subject between 1615 and 1625.[16] Summarizing the views of Goclenius, Roberti sought to enlist van Helmont to his side,[17] arguing that such a cure could only be the product of satanic forces.

Van Helmont would have been well advised to have refrained from entering this debate, but his reply appeared in 1621[18] and it provided little comfort to either party. He argued that Goclenius had greatly oversimplified the problem in arguing that the cure was natural, and also that Roberti had added nothing at all of value by asserting that the method was satanic. Indeed, this only confirmed his opinion that "Nature . . . called not Divines to be her Interpreters: but desired Physitians only for her Sons; and indeed, such only, who being instructed by the Art of the Fire doe examine the Properties of things . . ."[19] In short, the theologian had no right to meddle in the study of nature or its forces: ". . . let the Divine enquire concerning God, but the Naturalist concerning Nature."[20]

For van Helmont this cure was not diabolical because all true remedies must proceed from God. Here also he accepted the macrocosm-microcosm analogy and the three principles. He praised Paracelsus who

> hath disclosed Magnetism, unknown to Antiquity and in the room of that natural Study which is barrenly taught up and down in the Schooles, hath brought to us another real one; which by the Resolution, and Composition of Bodies is made probably to our hands, and far more plentiful in Knowledge; that from thence he hath rather by a just title, snatch'd away the Denomination of the Monarch of Secrets, from all that went before him . . .[21]

---

[16]The most convenient collection of texts relating to the weapon-salve is the *Theatrum Sympatheticum auctum* . . . (Nuremberg: J. A. Endten, 1662). Included are tracts by Digby, Goclenius, Roberti, van Helmont, Fludd, Kircher, Sennert, Conring, Fracastor, and many others.

[17]Van Helmont, "De magnetica" (sect. 1), *Ortus*, p. 748; *Oriatrike*, p. 759.

[18]J. B. van Helmont, *Disputatio de magnetica vulnerum naturali et legitima curatione, contra R. P. Joannem Roberti* (Paris: Victor Leroy, 1621). References will be made to the texts in the *Ortus* and the *Oriatrike*.

[19]"De magnetica" (sec. 8), *Ortus*, p. 750; *Oriatrike*, p. 761.

[20]*Ibid.*

[21]"De magnetica" (sect. 53), *Ortus*, p. 759; *Oriatrike*, p. 771.

Roberti must have been alarmed at van Helmont's definition of magic as "the most profound inbred knowledge of things"[22] and at his claim that the key to its revival might be found in the Cabala.[23] For van Helmont even the power of sacred relics could rightly be compared to natural magic.[24] Surely he believed that the action of the weapon-salve was natural magic as well.[25]

Van Helmont's work on the magnetic cure of wounds was denounced by the Medical Faculty at Louvain in 1623[26] and the following year he offended the influential physician, Henry van Heer, through his rejection of the latter's explanation of the origin of springs.[27] Further examination of his opinions in the weapon-salve tract led to action by the Spanish Inquisition, with the resultant charge of heresy and magic.[28] Censured not only by the doctors at Louvain, but also Cologne and Lyons, he acknowledged his errors in 1627 and again in 1630. Repeated examinations of his views led to his arrest in 1634 after he was accused of being a follower of Paracelsus and an adherent of the Chemical Philosophy, which had spread a "Cimmerian darkness" over the world.[29]

---

[22]"De magnetica" (sect. 126), *Ortus*, p. 771; *Oriatrike* (sect. 125), p. 784.

[23]"De magnetica" (sect. 106), *Ortus*, p. 769; *Oriatrike* (sect. 105), p. 781.

[24]"De magnetica" (sect. 46), *Ortus*, p. 757; *Oriatrike*, p. 769.

[25]"De magnetica" (sect. 106), *Ortus*, p. 769; *Oriatrike* (sect. 105), p. 781.

[26]On the prosecution of van Helmont, see Nève de Mévergnies, *Jean-Baptiste van Helmont, Philosophe par le feu* (Paris: Libraire E. Droz, 1935, pp. 123–43. See also C. Broeckx, *Commentaire de J. B. van Helmont sur le premier livre du Régime d'Hippocrate: Peri diaites* (Antwerp, 1849); C. Broeckx, *Interrogatoires de J. B. van Helmont sur le magnétisme animal* (Anvers: Buschmann, 1856); Willem Rommelaere, *Études sur J.-B. Van Helmont* (Brussels: Manceaux, 1868); and A. J. J. Vandevelde, "Helmontiana," 5 parts in *Verslagen en Mededeelingen. K. Vlaamsche Academie voor Taal-en Letterkunde*, Pt. 1 (1929), pp. 453–76; Pt. 2 (1929), pp. 715–37; Pt. 3 (1929), pp. 857–79; Pt. 4 (1932), pp. 109–22; Pt. 5 (1936), pp. 339–87.

[27]J. B. van Helmont, *Paradoxa de aquis Spadanis* (Leodii, 1624) and *Supplementum de Spadanis fontibus* (Leodii: L. Streel, 1624). These texts are to be found in the *Ortus*, pp. 685–99 and the *Oriatrike*, pp. 687–703. Henricus van Heer's *Spadacrene, hoc est fons Spadanus . . .* (Leodii: Arn. de Corswaremia, 1614) went through many editions in the seventeenth century and is of interest because of the author's concern with analytical procedures. An interest in color indicators is evident as early as the 1630 edition and later editions illustrate crystal forms. For van Helmont on the origin of springs, see van Helmont, "De Spadanis fontibus, Paradoxum secundum" (sects. 9–10), *Ortus*, p. 689; *Oriatrike*, p. 692.

[28]Here see Nève de Mévergnies, *Van Helmont*, pp. 123–30 and Marin Mersenne, *Correspondance du P. Marin Mersenne, Religieux Minime*, pub. Mme. Paul Tannery, ed. Cornelis de Waard with the collaboration of René Pintard [12 vols. (1617–1643), Paris: P. U. F. Centre National de la Recherche Scientifique, 1932–1971], *2*, pp. 499n, 589n.

[29]Nève de Mévergnies, *Van Helmont*, pp. 133ff.

Imprisoned at first in the archbishopric jail, van Helmont was shortly thereafter transferred to the Minorite convent in Brussels.[30] Forced to repudiate his errors once more, he was finally permitted to live at home, at first under house arrest. It is most likely that at this time he wrote the bulk of his works. During the winter of 1644 he became ill with pleurisy. Realizing the serious nature of his disease, he requested his son, Franciscus Mercurius, to take charge of his writings.[31] Shortly afterward he died, on December 30th, 1644.

THE "NEW PHILOSOPHY" OF VAN HELMONT

The arrangement of his father's work by Franciscus Mercurius van Helmont was to help make it one of the most influential medical works of the seventeenth century.[32] Published first in Latin in 1648 in an edition of more than 1,100 pages, it was reprinted in five other editions down to 1707. Translations also appeared in English (1662, 1664), French (1670, 1671), and German (1683), while a Flemish compilation probably prepared by van Helmont the elder appeared in 1659 and 1660. In addition there were numerous printings of individual works and small collections.

True, the organization of the work suffers unavoidably because of the fact that these works were not planned originally as a whole. Nevertheless, it is evident that van Helmont had in mind a total philosophy of nature–not simply a reformed medicine.

> For it was needful, that in composing a new Philosophy, I should break down almost all things that have been delivered by those that went before, and many things ought to be set in good order, and restored, which every one will not receive with a like acceptance.[33]

The natural philosophy of Aristotle was useless, for he had been totally ignorant of nature. It was his emphasis on logic that formed the basis of his "mathematical" method, which had led nowhere.

---

[30] *Ibid.*, pp. 135–43.

[31] Van Helmont, "Amico lectori" (Franciscus Mercurius van Helmont), *Ortus*, sig. $^{xx}1^{v}$; *Oriatrike*, sig. $b2^{v}$.

[32] For a detailed description of the various editions of the works of van Helmont, see Pagel, *Van Helmont*, pp. 209–14.

[33] Van Helmont, "Logica inutilis," *Ortus*, p. 41; *Oriatrike*, p. 37.

> . . . let the Schooles tell me, what Science *Logick* hath ever brought to light? whether happily Geometry? Musick? making of Glasse? Printing? Husbandry? Medicine? drawing or conducting of Water? of Mineralls? of Warring? of Arithmetique? of Building? or any profitable Science? verily none.[34]

For Aristotle, Nature "*is the Principle, or beginning of motion, as also of rest in Bodies, in whom it is in by it self, and not by accident.*"[35] The eight books of *Physics* of Aristotle present us with "dreams" that "suppose a matter, or impossible corporeity or bodyliness, with Mathematical abstraction, for the principle, prop, and seminary of nature."[36] How different this is from the contemporary move toward mathematical abstraction in the world of Galileo and other physicists! In a sense, van Helmont was proposing a far more revolutionary philosophy, one that would divorce mathematical logic and mathematical abstraction from the interpretation of nature.

Rather, "*Nature is the Command of God, whereby a thing is that which it is, and doth that which it is commanded to do or act.* This is a Christian definition, taken out of the holy Scriptures."[37] Van Helmont would not have found Galileo's study of local motion of much interest. Rather, he was concerned with the involuntary motions of the body, such as the beating of the heart. In these cases no contact with a mover was required. If the Aristotelians were correct in their belief that motion occurred only through contact with a mover, this could only mean that God as the Prime Mover was immobile himself.[38] Confining the Creator in this or any other way was unthinkable.

The authority of Holy Scripture was no less important to van Helmont than it had been to others in the Paracelsian tradition. Genesis I informed its readers that plants had been created prior to the stars. Animals and men, created later, could not be subject to the stars if plants were not. For this reason–and in contrast to his earlier views expressed in the weapon-salve tract–van Helmont now rejected the concept of man as a microcosm. It is "a phantastical, hypochondriacal, and mad thing to have brought all the properties, and species of the Universe into man, and the art of healing."[39] To be sure, there

---

[34] *Ibid.*, *Ortus*, p. 45; *Oriatrike*, p. 41.

[35] "Physica Aristotelis et Galeni ignara," *Ortus*, p. 46; *Oriatrike*, p. 42.

[36] *Ibid.*, *Ortus*, p. 49; *Oriatrike*, p. 45.

[37] *Ibid.*, *Ortus*, p. 46; *Oriatrike*, p. 42.

[38] "Blas Humanum," *Ortus*, pp. 179–80; *Oriatrike*, p. 176.

[39] "Scabies & ulcera scholarum" (sec. 33), *Ortus*, p. 328; *Oriatrike*, p. 323.

was an evident relationship between man and the stars, but this was at best an indirect influence because "the name therefore of Microcosm or little World is Poetical, heathenish, and metaphorical, but not natural or true."[40] Indeed, "I being a Christian could not admit of Microcosmicall Dreams, as they have been delivered by *Paracelsus*."[41]

Genesis was also to be a determinant for element theory. Here the all-important first chapter does not refer to the creation of fire.

> Therefore I deny that God created four Elements; because not the fire, the fourth. And therefore it is vain that the fire doth materially concurre unto the mixture of bodies. Therefore the fourfold kinde of Elements, Qualities, Temperaments or Complexions, and also the foundation of Diseases falls to the ground.[42]

With one reference to Genesis van Helmont dispensed with the basis of all ancient medicine and natural philosophy.

But were the Paracelsian three principles any better? Here he turned to the use of fire and heat as analytic agents.[43] It is true that they separate many things into three fractions that may be called "salt," "sulphur," and "mercury." But in fact, he argued, these three are not elemental at all. Rather, "*as by a Trans-mutation made by the Fire, they are there generated, as it were new Beings, and there is made that, which was not before*."[44] Specific examples are given to show that different results may be obtained from different applications of heat, while others show that some things can never be decomposed by heat or fire. In short, although the use of the terms Salt, Sulphur and Mercury might be useful, no one should think that they are true elements. This section was clearly the basis of Robert Boyle's later work, *The Sceptical Chymist* (1661).[45]

What are the true elements? In Genesis there is no mention of the creation of water and air that are included in Heaven. "Therefore I call these two

---

[40] *Ibid.*

[41] "Inventio Tartari, in Morbis Temeraria," *Ortus*, p. 243; *Oriatrike*, p. 239.

[42] "Elementa," *Ortus*, p. 53; *Oriatrike*, p. 48.

[43] I have discussed this at some length in "Fire Analysis and the Elements in the Sixteenth and the Seventeenth Centuries," *Annals of Science*, *23* (1967), 127–47.

[44] "Tria prima" (sec. 47), *Ortus*, p. 405; *Oriatrike*, p. 408.

[45] The connection with Robert Boyle and *The Sceptical Chymist* is made in Debus, "Fire Analysis." See n. 43.

Elements Primigeniall, or first-born, in respect of the Earth."[46] Still, van Helmont had a special interest in the elemental nature of water and he supported his view through the well-known willow tree experiment.[47] Here he planted a small willow tree in a weighed amount of earth and added nothing but water. At the end of five years the earth weighed the same, but the tree had gained 164 pounds that could only be accounted for by the water. There seemed little doubt that water was the origin of this "earth" and was the true *prima materia.*

Not only had the foundations of ancient philosophy and medicine been denied, but it now seemed that the basic principles of Paracelsian philosophy were to be rejected as well. Once again the answer was to be found in the proper use of chemistry.

> I praise my bountiful God, who hath called me into the Art of the fire, out of the dregs of other professions. For truly, Chymistry, hath its principles not gotten by discourses, but those which are known by nature, and evident by the fire: and it prepares the understanding to pierce the secrets of nature, and causeth a further searching out in nature, than all other Sciences being put together: and it pierceth even unto the utmost depths of real truth.[48]

This confidence in the significance of chemical procedures as a key to the understanding of nature in no way conflicted with van Helmont's belief in transmutatory alchemy, which he believed he had carried out himself with a sample of the Philosopher's Stone.[49] Nor did he think that the traditional secrecy of this art need be abandoned.

> For nothing in Alchymical things is written to that intent that they may be promiscuously understood by all, but onely, that they may not be understood: And that thing, Chymistry hath alwayes observed

[46]"Elementa" (sects. 7, 8), *Ortus*, p. 53; *Oriatrike*, p. 48.

[47]"Complexionum atque mistionum elementalium figmentum" (sec. 30), *Ortus*, pp. 108–9; *Oriatrike*, p. 109. The experiment has a long history and was described by Nicholas Cusanus in *The Idiot* [see Nicholas de Cusa, *Unity and Reform: Selected Writings of Nicholas de Cusa*, ed. John P. Dolan (South Bend, Ind.: University of Notre Dame Press, 1962), pp. 250–51]. See also H. M. Howe, "A Root of van Helmont's Tree," *Isis*, *56* (1965), 408–19.

[48]Van Helmont, "Pharmacopolium ac dispensatorium modernum" (sect. 32), *Ortus.*, p. 463; *Oriatrike*, p. 462.

[49]Van Helmont referred to his belief in transmutation on more than one occasion. A circumstantial account is given in the tract "Vita aeterna," *Ortus*, p. 743; *Oriatrike*, pp. 751–52.

> singular to it, before other Disciplines, by the Command of God; least Roses should be spread before Men, and Swine: For our Writings are in stead of Exhortations, that every one may profit by his own Labours, as much as shall be indulged him from above.[50]

This traditional endorsement of alchemical secrecy appealed to many. Those who saw in van Helmont a true adept could also cite his personal account of the transmutation of mercury to gold with a portion of the Philosopher's Stone that had been given to him.[51] In addition, his belief in the existence of a universal solvent, the *alkahest*,[52] and a universal remedy for all diseases[53] resulted in many disciples.

But van Helmont's chemistry went beyond this. We have already noted his quantitative approach in the growth of the willow tree. In a series of letters to Marin Mersenne in 1630–1631, he repeatedly referred to the use of quantification in his chemical laboratory. He compared the specific weights of metals[54] and also recommended the weight comparison of equal volumes of urine as a means of improving traditional uroscopy. He was concerned with the accurate determination of heat and he used the primitive barometer/thermometer, which had only recently been described by several authors, including Robert Fludd. More importantly, he accepted the principles of conservation of weight and the indestructibility of matter. And although his explanation was faulty, he agreed with the mechanists that Aristotle was wrong in denying the possibility of a vacuum.[55]

Along with most other natural philosophers of his period, van Helmont was a vitalist. For him the growth of metals and minerals in the earth was no less certain than the growth of plants or of animals. In all cases the growth process was due to a seed acting in an appropriate matrix by means of an internal force or archeus. In his detailed description of mineral growth[56]

---

[50]"Puerilis humoristarum vindicta," *Ortus*, p. 523 (Incorrectly pagniated 524); *Oriatrike*, p. 523.

[51]"Vita aeterna," *Ortus*, p. 743; *Oriatrike*, pp. 751–52.

[52]"Explicatio aliquot verborum artis," *Opuscula*, sig. A4^v; *Oriatrike*, sig. Nnnnn4^v.

[53]"Butler," *Ortus*, p. 587; *Oriatrike*, p. 587.

[54]I have discussed these and the following examples indicating van Helmont's interest in quantification in *The Chemical Philosophy*, *2*, pp. 327–29.

[55]*Ibid.*, pp. 329–34. Van Helmont discusses the vacuum in "Vacuum naturae," *Ortus*, pp. 83–87; *Oriatrike*, pp. 81–87.

[56]For van Helmont on mineral growth, see Debus, *Chemical Philosophy*, *2*, pp. 339–43. Mineral growth in the earth is described by van Helmont in his "Magnum oportet" (sects. 39, 40), *Ortus*, p. 157; *Oriatrike*, pp. 155–56.

he referred to a proto-mineral, *Bur*, which was widely accepted and even made the subject of investigation by the early Fellows of the Royal Society of London.

HELMONTIAN MEDICINE

Clearly van Helmont sought both to destroy the foundations of Aristotelian natural philosophy and to establish a basis for a *nova Philosophia*. And yet, his philosophy was always of most interest to physicians. Like others before him, van Helmont cited Ecclesiasticus, chapter 38, to confirm that only God could create a true physician.[57] The fact that the medicine taught at the universities was that of the Greeks, the Jews, and the Arabs, should be enough, he thought, to condemn it in the eyes of Christians.[58] Such authorities had sought only personal gain in this art and could never have been called to it by God.

But in addition to these shady origins there was the fact that the humoral medicine of the Galenists was invalid because the quaternary of elements upon which it was based did not exist.[59] Van Helmont was also able to apply his concept of growth processes in the three kingdoms of nature to disease theory. Seeds of disease might enter the body and lodge in an organ, where the archei of the disease and that of the afflicted organ would battle for control.[60] Although this concept of the archeus is reminiscent of that of Paracelsus, it is here more defined and surely the understanding of disease as localized in the body was different from the Galenic concept of health as the correct balance of humors.

Chemistry is an ever recurring theme in van Helmont's discussion. Convinced of the existence of a material vital spirit, he sought to isolate it through the distillation of arterial blood.[61] His product was a salt that was similar to that obtained from urine, but the two salts differed in that "the Spirit of the

---

[57]This is referred to frequently by van Helmont, but in most detail in the "De Lithiasi" (Philiatro lectori), *Opuscula*, p. 4; *Oriatrike*, sig. Nnnnn2^v.

[58]*Ibid.*, *Opuscula*, p. 5; *Oriatrike*, sig. Nnnnn3^r.

[59]"Scholarum humoristarum passiva deceptio ignorantia" (Ch. 1, sec. 3), *Opuscula*, p. 70; *Oriatrike*, p. 1017. See also the "Natura contrarium nescia" (sect. 1), *Ortus*, p. 164; *Oriatrike*, p. 161.

[60]I have discussed van Helmont's theory of disease in *The Chemical Philosophy*, *2*, pp. 359–362. Here I follow Walter Pagel's work closely. In addition to his book on van Helmont, see his major article, "Van Helmont's Concept of Disease–To Be or Not to Be? The Influence of Paracelsus," *Bull. Hist. Med.*, *46* (1972), 419–54.

[61]"Aura vitalis," *Ortus*, p. 726; *Oriatrike*, p. 733.

Salt of Venal Blood cureth the Falling-sickness, but the Spirit of the Salt of Urine not so." And more, "the vitall Spirit is Salt, not soure . . . being nearer to the Spirit of Urine than to Salt-peter . . ."[62] "Therefore the vitall Spirit *is arterial bloud resolved by the Ferment of the heart, into a salt Air, and enlightned by life.*"[63] Because of this, bloodletting could not be condoned, because the result would be to lessen the vital spirit of the body.[64]

If chemical processes might lead to the isolation of the vital spirit, they might also be employed as a dissolvent for bladder stones or urinary calculi.[65] In this case, acids proved to be effective in their action on the stones, but they could not be taken by mouth because of the alteration they underwent in the digestive process. On the other hand, their direct application through the urinary tract was too painful. It was a dilemma that would continue to trouble physicians long after the death of van Helmont.

Although van Helmont agreed with other Paracelsians that contraries did not cure, he did not adhere to the dictum that like cures like. Rather, he believed in a search for specific medical powers and inherent virtues that might be effective in the cure of disease.[66] And although van Helmont did not present the reader with a chemical pharmacopoeia, he surely supported the use of chemically prepared medicines. He was well aware "*that the more choice* Physitians *at this day, do not despise Chimical remedies, the which, their bookes do lately testifie.*"[67]

There were those who reproached the chemists for preparing their own medicines at the furnace, but what of the atrocious substances in the traditional materia medica? "Is it happily a more glorious thing for the Galenist rout to have viewed stinking dung, and to have stirred it with a stick, than for us to have handled Furnaces, vessells, and coales."[68] And while he was willing to grant that "all kind of Knaves have most licentiously thrust themselves

---

[62]*Ibid.*, *Ortus*, p. 727; *Oriatrike*, p. 734.

[63]*Ibid.*, *Ortus*, p. 728; *Oriatrike*, p. 734.

[64]"De febribus" (Ch. 4), *Opuscula*, pp. 17–25; *Oriatrike*, pp. 949–57. See P. H. Niebyl, "Galen, Van Helmont and Blood Letting," *Science, Medicine and Society in the Renaissance*, ed. Allen G. Debus, 2 vols. (New York: Science History Publications; London: Heinemann, 1972), *2*, pp. 13–23.

[65]"De Lithiasi" (Ch. 7, sect 28), *Opuscula*, p. 65; *Oriatrike*, p. 884. Van Helmont's work led him to develop a less painful catheter for urinary injection. See *ibid.* (Ch. 7, sect 34), *Opuscula*, pp. 67–68; *Oriatrike*, p. 407.

[66]"Tria prima chymicorum principia" (sects. 42, 43), *Ortus*, p. 404; *Oriatrike*, p. 407.

[67]"De febribus" (Ch. 15, sect. 3), *Opuscula*, p. 54; *Oriatrike*, p. 990.

[68]*Ibid.*

into Chymistry, no less than Medicinal Affairs, and that a various destruction doth thereby arise unto mankind; . . . these things do not defame honest men."[69] In fact, chemistry alters corrosive substances that are manifest poisons into substances sweeter than sugar.

> For truly Scarwort, Frogwort, *Apium risus*, &c. do forthwith in distilling lay aside their embladdering power, even as the juyce of Citron doth its sharpness, Water-Pepper its Acrimony, &c. Neither is that of concernment, that Chymical Medicines are to be administered in a small Dose: For that accuseth them not of poysonsomness, but of the highest perfection of acting . . .
>
> The virtues of a Chymical remedy are narrowed in a small quantity, under which they are pleasing while as all things have regard in their own proportions, unto the strength and necessities of the sick.[70]

Still, it was necessary to warn against the internal use of numerous remedies: as to "Arsenick and Orpiment, &c. How much soever they may be fixed, and dulcified or made sweet; yet they are never to be taken inwardly, however others shall otherwise perswade."[71]

In short, Galenic physicians have built up a complex armory of mixtures, and then mixtures, and then mixtures of mixtures so that they

> give sometimes the hotch-potch of a thousand simples to the sick, to drink, that if one thing help not, at least wise another may help . . . This the preparation of medicine at this day, from which, how far the Philosophy of Chymistry differs, they indeed have known, who even but from the entrance, have saluted the same; but unskilfull haters only are ignorant thereof."[72]

Van Helmont's chemical interests led him ever more toward an interest in acids and alkalis. He wrote that he had learned that the digestive fluid of the stomach is an acid while he was a boy. At that time

---

[69] *Ibid.*, (Ch. 15, sect. 4), *Opuscula*, p. 54; *Oriatrike*, p. 990.

[70] *Ibid.*, (Ch. 15, sect. 4), *Opuscula*, p. 54; *Oriatrike*, pp. 990–91.

[71] "Pharmacopolium ac dispensatorium modernum" (sect. 52), *Ortus*, p. 468; *Oriatrike*, p. 466. In the "Pharmacopolium," van Helmont discusses the dangers as well as the values of various medicines.

[72] "De febribus" (Ch. 15, sec. 9), *Opuscula*, pp. 55–56; *Oriatrike*, p. 992.

> I nourished Sparrows; [and] I oft-times thrust out my tongue, which the Sparrow laid hold of by biting, and endeavoured to swallow to himself, and then I perceived a great sharpness to be in the throat of the Sparrow, whence from that time I knew why they are so devouring and digesting.
>
> And then I saw that the sharp distilled Liquor of *Sulphur* had seasoned my Glove, and that it did presently resolve it into a juice, in the part which it had moistned; which thing confirmed to me a young Beginner, that meats are transchanged by a sharp or soure thing, and so that a ferment doth inhabit in the stomack, which should change all things cast into it, although sweet, presently into a sowreness.[73]

This observation led van Helmont to reject the traditional explanation of digestion given by contemporary Galenists. They held that digestion occurred in three steps. As van Helmont presented their views, the first step, centered in the stomach, was dependent on heat: "*meat and drink are by the force of heat transchanged into a liquor; the which, by reason of its likeness to Barley Cream, they have called Chyle.*"[74] A second stage occurred in the liver, where the urine was separated from the venal blood. Finally, the blood from the veins was distributed for the nourishment of the solid parts of the body. "Therefore in this alimentary humour, is bestowed the labour of the third and highest digestion."[75]

In his rejection of the Galenic system, van Helmont described a six-stage digestive process that was to be understood largely in chemical terms through the action of acids, alkalis, salts, and ferments. The first digestion was centered in the stomach.[76] Here the spleen added a sour (acidic) element to the food and changed it to a cream through fermentation. The second digestion occurred in the small intestine due to the ferment furnished by the bile.[77] This caused a separation of the thicker from the more watery parts. In this way the urine was removed prior to sanguification. The mesentery veins became the third digestive center; there a ferment from the liver produced the venous

[73]"Calor efficienter" (sects. 23, 24), *Ortus*, p. 204; *Oriatrike*, p. 201. Pagel refers to the Flemish and German editions and concludes that van Helmont identified the acid as hydrochloric. "Van Helmont's Ideas on Gastric Digestion and the Gastric Acid," *Bull. Hist. Med.*, 30 (1956), 524–36.

[74]"Triplex scholarum digestio," *Ortus*, pp. 207–8 (207); *Oriatrike*, pp. 203–4 (203).

[75]*Ibid.*, *Ortus.*, p. 207; *Oriatrike*, p. 204.

[76]"Sextuplex digestio alimenti humani," *Ortus*, pp. 208–25 (213); *Oriatrike*, pp. 205–21 (210).

[77]*Ibid.*

blood.[78] The fourth digestion then occurred in the heart, where venous blood was changed into arterial blood.[79] And the fifth took place in the left ventricle, where the arterial blood was transcharged "into the vital spirit of an Archeus" and distributed through the arteries.[80] "At length, the sixth and last Digestion is perfected in all the particular Kitchins of the Members: And there are as many stomacks, as there are members nourishable."[81] The number of digestions had a similarity to the days of the Creation. "Nature therefore hath with me, constituted six vital digestions; But in the seventh number she her self resteth."[82]

Van Helmont's work on digestion led him to an understanding of osmosis[83] as well as the observation that neutralization occurs in the small intestine through the action of the bile.

> The Cream sliding out of the *Pylorus* or neather mouth of the Stomack, unto the Duodenum, being straight-way snatched within the Sphear of activity, by the in-breathing of the Gaul, doth exchange its sourness into a Salt, and its more watery part is made severable from its more pure of un-mixt part, which is drawn by the Reins.[84]

Similarly, he noted that "in every Wound, A Tartnesse or Acidity, the Betokener and Companion of all putrefaction in the flesh, doth arise: the which *Alcalies* do easily sup up into themselves and consume."[85] The concept of neutrality is set forth clearly in his discussion of the action of fire on substances, where he wrote that "now and then, the things themselves together with their adjuncts, are diversely transcharged by the Fire, and become neutral."[86] This use of acid-base explanations was to influence later seventeenth-century authors who viewed chemistry as a true basis of understanding physiological processes.

---

[78] *Ibid.*

[79] *Ibid., Ortus*, p. 221; *Oriatrike*, p. 218.

[80] *Ibid.*

[81] *Ibid., Ortus*, p. 222; *Oriatrike*, p. 219.

[82] *Ibid., Ortus*, p. 225; *Oriatrike*, p. 221.

[83] I have discussed the references pertinent to osmosis in *The Chemical Philosophy*, *2*, pp. 370–71.

[84] Van Helmont, "Sextuplex digestio" (sect. 56), *Ortus*, p. 220; *Oriatrike*, p. 217.

[85] "De Lithiasi" (Ch. 5, sect. 17), *Opuscula*, p. 45; *Oriatrike*, p. 863.

[86] "Potestas medicaminum" (sects. 36, 37), *Ortus*, p. 479; *Oriatrike*, pp. 477–78.

Allegorical frontispiece to the German translation of the works of van Helmont [*Aufgang der Artzney-Kunst* (Sulzbach, 1683)] illustrating his dream at the tomb of the true medicine. Note the light of knowledge and understanding falling on van Helmont alone. The figures at the top indicate the importance of the mining of minerals and their chemical use in medicine. Department of Special Collections, the University of Chicago.

Van Helmont was convinced that he had gone beyond all those before him in his study of nature and medicine. At the opening of his *Tumulus pestis*, he recounted a dream in which he had found himself in the vaults outside of a city, an area of rubble and decay with poisonous gases.[87] It was all clothed in darkness and the only life was a company of bats. It was the image of eternal death, which he recognized as the covering that must be breached to attain the Sepulchre of Truth. But the "unhappy birds of Night suffer not others to enter."[88] Indeed, although they "do not know Truth themselves, . . . [they] . . . demand others prostrate themselves to them."[89] Undaunted by this "I endeavoured by the unwearied pains and charges of forty years, to break the rocky stones asunder . . . that light may flow in from heaven, and that the Night-birds which presume to keep the Keys of Sciences . . . may vanish away . . ."[90] Galen earlier had "entered into the Vaults with a slender Lamp." He had stumbled and almost fallen. "Therefore, his Oyl being lavishly spent, he returned to his own, and told many things confusedly, concerning the Sepulchres, which he had not perceived, nor known, nor believed, although he had seen them."[91]

Then came Avicenna, who entered not much deeper than Galen. Dashing his foot against a stone, he fell down. Nevertheless, on his return, he boasted that he had seen far more than his predecessors.[92]

> At length *Paracelsus* having entered with a great Torch, fastened a small cord to the wall, about his first paces, which he might follow as a Companion, and Reducer of the wayes; he aspiring to pierce whither the footsteps of mortals had not yet taken their journey. The rout of Birds is presently amazed at so great a light, it thinks that Prometheus had entred; it does not nor was able to extenguish the Torch. . . . This man seeth very many Monuments, he is long and freely enlarged, he fills the entries with smoak, and while he is intentive, as a greedy devourer of truth, his strength fails, his Torch falls, his light is extinguished in the middle of his course, and he is as it were choaked with fumes.[93]

---

[87]His dream is recounted in the "Tumulus pestis," *Opuscula*, pp. 5–8; *Oriatrike*, pp. 1073–75.

[88]*Ibid.*, *Opuscula*, p. 5; *Oriatrike*, p. 1074.

[89]*Ibid.*

[90]*Ibid.*, *Opuscula*, p. 6; *Oriatrike*, p. 1074.

[91]*Ibid.*, *Opuscula*, p. 6; *Oriatrike*, pp. 1074–75.

[92]*Ibid.*, *Opuscula*, p. 6; *Oriatrike*, p. 1075.

[93]*Ibid.*

Detail of the dream engraving, showing the failure of Galen, Avicenna, and Theophrastus Paracelsus in comparison with van Helmont. Department of Special Collections, the University of Chicago.

Then followed van Helmont:

> I a poor miserable man have at length entred with the least light of a Lanthorn; and that nothing might hinder, and that nothing might detain my hand from the work, I indeed refused a Rope, and hung my Lanthorn at my girdle, but a Crook followed at my back, making a path the rule of my return: Therefore I insisting only in my own footsteps, I there saw far other things than the foregoing company of Ancestors had described.[94]

And yet, with the crowd of bats against him, he returned with no more than his predecessors. He had, however, learned that by trusting wholly to human aids

[94] *Ibid.*, *Opuscula*, p. 7; *Oriatrike*, p. 1075.

we would never uncover truth. In time he recognized that these truths were the gift of the Lord given freely to those chosen to be physicians. This was a story with a moral that appealed to many in the mid-seventeenth century.

THE EARLY REACTION TO VAN HELMONT: FONTAINE AND GREMBS

At the death of van Helmont in 1644, chemistry was still a subject of heated debate in European medical circles. The main subject of contention was surely the use of the strong metallic and mineral remedies, but the relentless attack by the Paracelsians on the ancient authorities and the call for a new observational chemical philosophy pushed Aristotelians and Galenists ever more toward sharp opposition.

Writing to Charles Spon in Lyon on April 16, 1645, that long-time enemy of all chemists, Gui Patin,[95] took heart in hearing of the death of van Helmont, whom he called an evil Flemish rogue. Patin considered all that van Helmont had written worthless–and he had read it all! He objected to the fact that van Helmont thought that physicians had to know the secrets of chemistry and the work of empirics. In addition, Patin complained that van Helmont had rejected bloodletting, for the lack of which he had died raving mad.[96]

Ten years later Patin commented on the 1655 edition of van Helmont's *Opera*, which he considered of no value. It seemed to him that the book was useless for anyone, and he referred specifically to scholars and physicians.[97] In 1661 he wrote again to Spon when a French translation of van Helmont was being considered: "I have nothing good to say of a translation of van Helmont: it is a stupid book and the author a malicious knave; in French this book would not have a large sale."[98] Indeed, it would seem that Patin's spirits were buoyed only when chemists were attacked in print. Thus, in a letter to the younger doctor Belin at Troyes dated the 24th of March 1657, he noted a newly published book at Lyon entitled *Medicina anti-emetica* [sic] in

[95]I have discussed the place of Gui Patin in the French chemical debates of the seventeenth century in *The French Paracelsians: The Chemical Challenge to Medical and Scientific Tradition in Early Modern France* (Cambridge University Press, 1991), pp. 91–96.

[96]Gui Patin, *Lettres . . . Nouvelle Édition augmentée de lettres inédites, précédée d'une notice Biographique . . . par J. -H. Reveillé-Parise* (3 vols., Paris: J. -B. Baillière, 1846), *1*, p. 355.

[97]*Ibid.*, *2*, p. 178.

[98]*Ibid.*, *2*, p. 461.

which the author, Gabriel Fontaine, refuted van Helmont, "the great impostor, unworthy of being refuted."[99]

This book had been written by Gabriel Fontaine (fl. 1655) whose father, Jacques (d. 1621), had written against Paracelsian element theory (1581) and magic (1613).[100] His son paid tribute to the refutation of Paracelsian thought by Thomas Erastus (1572–1574), but felt that a new refutation was required because of the persistence of this medical heresy. In his survey of the medical world of his day,[101] Fontaine noted first those who adhered to the true medical views of the ancients, particularly the followers of Hippocrates and Galen. To these he added the medieval physicians, Avicenna, Mesue, Rhazes, Serapion, and Averroes. A second group consisted of the chemists who emphasized the value of their art in the preparation of medicinal substances. Here again we find Arabic contributors as well as specific authors from the Latin West–Albertus Magnus, Arnald of Villanova, Ramon Lull, and more recently Guinter von Andernach and Johannes Hartmann. All of these may be referred to, with admiration, but of a different sort were Paracelsus and his followers such as Joseph Duchesne, Oswald Crollius, Claude Dariot, and Jean Baptiste van Helmont. To be sure, Andreas Libavius had vindicated alchemy from its Paracelsian errors, but these later chemists continued to pervert the true Hippocratic medicine. Paracelsus, he believed, had even used black magic in his medicine.[102]

Fontaine did not question the value of chemistry for the preparation of some medicines, but he did not approve of antimonial spirits or the sulphur of arsenic as medicines.[103] Further, he opposed many of the fundamental beliefs of the chemical philosophers. He affirmed the existence of the four elements and he rejected the Paracelsian principles: "Neither the heavens nor

[99] *Ibid.*, *1*, p. 225.

[100] On Fontaine, see Walter Pagel, *The Smiling Spleen: Paracelsianism in Storm and Stress* (Basel et al.: S. Karger, 1984), pp. 69–76 and Debus, *The French Paracelsians*, pp. 115–16.

[101] Gabriel Fontaine, *De veritate Hippocraticae Medicinae Firmissimis rationum & experimentorum momentis stabilita, & demonstrata: seu Medicina Antihermetica, In qua Dogmata Medica physiologica, pathologica, & therapeutica, contra Paracelsi, & Hermeticorum placita clarissmé promulgantur: non reiectis penitus Chjymicorum inuentis, ad Hippocraticam artem conferentibus. Adiectus est ad calcem generalis index rerum in hoc opere contentarum: necnon introductio ad methodum medendi; atque Apologeticon aduersus Van-helmont, vbi firmissimé demonstratur, quatuor humores Galenistarum non esse fictitios: caetera octauo abhinc folio recensentur* (Lyon: Philip Borde, Lavr. Arnaud, & Cl. Rigavd, 1657), sig. e4^r–e4^v.

[102] Pagel, *Smiling Spleen*, p. 70.

[103] Fontaine, p. 210.

The allegorical tree of Oswald Grembs [frontispiece to the *Arbor Integra et Ruinosa Hominis* (Frankfurt, 1657)] indicating the favorable and unfavorable archei affecting health, illness, and death. From the collection of the author.

the body of the elements are composed of the three principles."[104] He rejected the macrocosm-microcosm analogy and singled out van Helmont for his work on the weapon-salve that condoned sympathetic medicine.[105] Also important to Fontaine was the theory of cure. Here he rejected the Paracelsian belief that like cures like (*similia similibus*) and supported the Galenic concept of cure by contraries.[106] Fontaine's "Apologeticon Adversus Van-Helmont" concentrated on the existence of the four humors that were the fundament of the true medicine of the ancients.[107] The added stimulus given to the Paracelsian tradition by the work of van Helmont is evident in Fontaine's belief that his own book offered Galenists an updated work that would replace that of Erastus.

In contrast to the work of Fontaine, there was also published in 1657 the *Arbor Integra et Ruinosa Hominis* of Franciscus Oswaldus Grembs (fl. 1657–ca. 1682).[108] This book is a Helmontian treatise that served a purpose similar to that of the *Idea medicinae philosophicae* (1571) of Peter Severinus. Physician to Guidobald, the Archbishop of Salzburg, Grembs referred in his title to the archei, both those that contributed to health as well as those that were responsible for death and disease.[109] The work itself was divided into three books: the first on an understanding of nature and factors relating to the health of man, the second on the present ruinous state of humanity, and the third on the restitution of the latter through Paracelsian, that is, chemical, remedies.

The reader familiar with van Helmont will recognize many concepts to be found in Grembs. The elements are discussed one by one with the rejection of fire.[110] And following van Helmont, Grembs denied the transmutation of air into water, asserting that it is a constant and permanent element.[111] He discussed the Creation, semina, fermentation, the archei–all Helmontian topics. Then he proceeded to the generation of stones, minerals, metals, salts,

---

[104]*Ibid.*, p. 1.

[105]*Ibid.*, pp. 183–91.

[106]Pagel, *Smiling Spleen*, p. 71.

[107]Fontaine, "Apologeticon Adversus Van-Helmont" in *op. cit.*, pp. 245–309.

[108]This work was drawn to my attention by Walter Pagel shortly before his death.

[109]The symbolic plate is described in the "Emblematis Explicatio," Franciscus Oswaldus Grembs, *Arbor Integra et Ruinosa Hominis, Id est Tractatus Medicus Theorico Practicus* . . . (Frankfurt: Joh. Georg. Spörlin, 1657), sig. A1^r–A2^r.

[110]*Ibid.*, p. 5.

[111]*Ibid.*, p. 9.

vegetables, and animals. He followed van Helmont in his description of the acid digestion in the stomach, specifically rejecting the Galenic belief that it was due to heat.[112] Rather, the "transmutation" of the food in the stomach was ascribed to a fermentation generated by a "hungry" acid. From this point Grembs followed the Helmontian six-fold digestive process. It is evident that Grembs was well read, but it is notable that like van Helmont he made no reference to William Harvey or the circulation of the blood.

The second book on the ruined state of man centered on specific diseases, but the third book offered hope through a discussion of chemical remedies. In a section on the lengthening of life, Grembs included a discussion of tobacco, which he condemned.[113] The liquor made from tobacco was shown to be poisonous and the fumes from tobacco affected the brain and interfered with conception. Smokers, he insisted, live short lives. Like van Helmont, Grembs rejected bloodletting as a medical practice,[114] while he affirmed that sexual abstinence would contribute to longevity.[115]

Grembs believed in a universal medicine that would cure all diseases[116] and he had a special interest in van Helmont's alkahest, which was claimed to be a solvent for all things. Above all, chemistry was a true key to nature because it taught not through logic, but through a knowledge of matter. It prepared the intellect to understand hidden aspects of nature because the chemist learned through the use of fire to reach the roots of all things. He could separate the fixed from that which was not fixed and he could reduce metals to their natural salts and sulphurs. Chemistry taught digestions through an understanding of ferments, and it opened substances, even the hardest and the most difficult, and showed how they might be converted into juices that could be medically compatible with our bodies.[117]

Grembs was aware of the enemies of chemistry who denied its truth and who insisted that remedies prepared in this way were poisonous and acted as violent purgatives.[118] Some even complained that it was indecorous to introduce coals into medicine through the fire. Such arguments were, however, of little account to Grembs, who praised the benefits brought to medicine by

[112] *Ibid.*, p. 80.

[113] *Ibid.*, pp. 465–67.

[114] *Ibid.*, p. 469.

[115] *Ibid.*, p. 487.

[116] *Ibid.*, p. 493.

[117] *Ibid.*, pp. 498-99.

[118] *Ibid.*, pp. 499ff.

iatrochemists such as Beguin, Mynsicht, Fabre, Anthony, van Helmont, Paracelsus, and Hartmann. However, like the Belgian he so admired, Grembs declined to give his readers a chemical pharmacopoeia.

CONCLUSION

The tracts of Fontaine and Grembs, both published in 1657, indicate the concern and interest of continental physicians that resulted from the publication of van Helmont's *Ortus medicinae*, which seemed to cap the new extensive canon of medical chemistry, in 1648. It is clear in both cases that although the chemistry associated with the preparation of the new remedies was important, there were further considerations that centered on traditional medical and philosophical beliefs. Although chemical descriptions of physiological processes were not yet predominant, this was to change with the studies of Franciscus de la Boë Sylvius at Leiden and Thomas Willis at Oxford.

CHAPTER 3

Chemistry, Anatomy and Physiology: Sylvius, Willis and Boyle

> *Let me . . . call in to my counsel and assistance,* Chymistry, *an Art which produceth things admirable and stupendious every day, and is certainly of principal use in the discovery of natural Mutations, and in some Particulars, if I may speak it, exceeding Nature her self; the most profitable and only necessary Means, for the constituting of Natural Science, and a Solid Body of Physick.*
>
> Marchamont Nedham quoting
> Franciscus de la Boë Sylvius, 1665[1]

The works of Gabriel Fontaine and Oswald Grembs reflect the early reaction to the Helmontian corpus that had appeared in a number of editions by the early 1650s. But interest in the revived chemical medicine of van Helmont was further fueled by the work of Franciscus de la

[1]Marchamont Nedham, *Medela Medicinae, A Plea for the Free Profession, and a Renovation of the Art of Physick, Out of the Noblest and most Authentick Writers. Shewing*
The Publick Advantage of its Liberty.
The Disadvantage that comes to the Publick by any Sort of Physicians, imposing upon the Studies and Practise of them.
The Alteration of Diseases from their Old State and Condition.
The Causes of that Alteration.
The Insufficiency and Uselessness of meer Scholastick Methods and Medicines, with a necessity of new.
Tending to the Rescue of Mankind from the Tyranny of Diseases; and of Physicians themselves, Pedantism of old Authors and present Dictators (London: Richard Lownds, 1665), p. 504.

Portrait of Franciscus de la Boë Sylvius (1614–1672). Courtesy of the National Library of Medicine.

Boë Sylvius (1614–1672) in the Netherlands and by a group of *virtuosi* at Oxford, particularly Thomas Willis (1621–1675). Here we meet with authors whose influence affected medical thought for half a century and more.

To Sylvius and Willis we may add Robert Boyle and Marchamont Nedham. At first glance neither would seem to be a major player in the field of medicine. But Boyle's earliest chemical works directly affected medical thought, while Marchamont Nedham, primarily a journalist, cited van Helmont, Willis, Sylvius, and Boyle repeatedly in his *Medela Medicinae* (1665), a work that called for medical reform on chemical lines. Widely read, this work was answered in short order by opponents who sought to uphold tradition and the authority of the Royal College of Physicians.

SYLVIUS AND THE CHEMISTRY OF THE BODY

Franciscus de la Boë, called Sylvius, was born in Hanau, Germany in 1614.[2] He studied medicine at Leiden from 1633 to 1635 and then at Wittenberg and Jena before taking his degree at Basel in 1637.[3] The following year he was back at Leiden giving lectures in which he defended the circulation of the blood. Although he was not the first to do so, his early acceptance of the circulation reflects his broad reading and understanding of the current literature. Sylvius also knew Descartes and he became a friend of Johann Rudolf Glauber when he moved to Amsterdam in 1640/1641. Surely this contact as well as his training at Jena, one of the first European universities to offer a course in chemistry, must have stimulated his interest in chemistry, which came to dominate his views on physiology.

Others were to suggest that Sylvius had borrowed his chemical concepts from van Helmont, but he denied this, stating that he had formulated his

---

[2]The only major study of Sylvius is in Dutch: E. D. Baumann, *François de la Boë Sylvius* (Leiden: E. J. Brill, 1949). Lester S. King has presented an insightful study of his work in *The Road to Medical Enlightenment 1650–1695* (London: Macdonald/NewYork: American Elsevier, 1970), pp. 93–112. Valuable for those interested in Sylvius's chemistry is J. R. Partington, *A History of Chemistry* (vol. 2, London: Macmillan, 1961), pp. 281–90. The short piece by G. A. Lindeboom in the *Dictionary of Scientific Biography* is disappointing, *13* (1976), pp. 222–23.

[3]Sylvius's lectures on anatomy, given at Leiden in 1640–1641 are printed in the *Opera Medica* (1st edition, 1679). Here I have used the second edition (Amsterdam: Daniel Elsevier and Abraham Wolfgang, 1680), pp. 875–94. He states (p. 891): "Quemadmodùm in pulmones amandati sanguinis pars minima in eorum abit nutricatum, maxima ad Cor revertitur, sic & qui in magnam arteriam in reliquum corpus propellitur sanguis per venas refluit, qui Circularis vocatur sanguinis motus."

views on the chemistry of the body early, and lectured on them in 1641, prior to the publication of van Helmont's works.[4] A distinction between the two may be easily made. Van Helmont sought a grand scheme of the world that was permeated with religious and hermetic thought. Although he differed from Paracelsus on many points, this Paracelsian influence continued on through his later writings. We search in vain, however, in the work of Sylvius for any trace of the mystical hermetic world view that so characterized the work of earlier chemical physicians. What we do find, however, is a description of disease and the body largely in chemical terms.

Sylvius became professor of medicine at Leiden in 1658 and there he was famous as a teacher and as a clinician. Among his students were Jan Swammerdam, Nicholas Steno, and Regnier De Graaf. Steno wrote that Sylvius presented his views in such a way that he stimulated others to investigate them.[5] This is particularly evident in DeGraaf's study of the pancreatic juice.[6]

Sylvius took an active role in the administration of the University of Leiden, serving as Vice-Chancellor in 1669–1670. But he also sought to develop and bring up to date medical teaching at the University, the subject of most interest to him personally.[7] Accordingly, when Anton Deusing (1612–1666) was being considered for an appointment by the medical faculty in 1666, Sylvius objected and threatened to resign unless he was appointed professor of chemistry and furnished with a properly equipped laboratory. Not wishing to lose Sylvius, the University Board acceded to his demands, noting that "nothing was lacking to make the distinction of the Medical Faculty complete

---

[4]This is discussed in Partington, *op. cit.*, p. 282; he considers the work of Sylvius superficial compared to that of van Helmont. On the other hand, José María López Piñero wrote that the work of Sylvius "was the first modern medical system, which confronted traditional Galenism in every sphere and endeavoured to integrate the novelties which had arisen over a period of more than a century . . ." "El Primer Sistema Medico Moderno: La Iatroquimica de la Segunda Mitad del Siglo XVII." "I. Posición histórica del movimiento. Silvio y la iatroquímica holandesa y alemana." *Medicina Española*, *67* (1972), 164–73 (166).

[5]Harold Moe, "Niels Stensen, When Steno Brought New Esteem to Glands," Jacob E. Poulsen and Egill Snorrason, eds., *Nicolaus Steno 1638–1686: A Re-consideration by Danish Scientists* (Gentofte, Denmark: Nordisk Insulinlaboratorium, 1986), pp. 51–96 (56).

[6]Regnier de Graaf's work on the pancreas was translated into English by a proponent of chemical medicine who also was to translate the work of Johann Rudolph Glauber. See Regnier de Graaf, *De succo Pancreatico: Or, a Physical and Anatomical Treatise of the Nature and Office of the Pancreatick Juice*. . . , trans. Christopher Packe, Med. Lond. (London: N. Brook, 1676).

[7]On Sylvius and his role in the introduction of chemistry to the University of Leiden, see Allen G. Debus, "Chemistry and the Universities in the Seventeenth Century," *Academiae Analecta* [*Klasse Wetenschappen*], *48* (1986), 15–33 (30).

but the preparation of medicaments in a chemical manner and the performance of experiments in the field of chemistry." And if Sylvius was not actually appointed Professor of Chemistry himself, at least a student of Glauber, Carel de Maets (1640–1690), was appointed three years later (1669) as an unsalaried docent.

Among the more important works of Sylvius are the *Disputationum medicorum decas* (1670), the *Praxeos medicae idea nova* (4 vols., 1671–1674), and the collected works, the *Opera medica*, which was published first in 1679 and frequently thereafter. Like other chemists of the seventeenth century, Sylvius rejected the idea that the ancients should be followed blindly. Indeed, he believed that the views of the medical authorities of antiquity were faulty and that they could not be relied upon.[8] For this reason current medical theory could not be relied upon either. Rather, we must turn to experience,[9] and here the key was to be found in chemistry,

> *an Art produceth things admirable and stupendious every day, and is certainly of principal use in the discovery of natural Mutations, and in some Particulars, if I may speak it, exceeding Nature her self; the most profitable and only necessary Means, for the constituting of Natural Science, and a solid Body of Physick.*[10]

Thus chemistry might be employed to explain thirst as the result of "*a Salt Humor distilling from the Head to the Gullet*," which there produced the sensation of thirst, while digestion and hunger were attributed to fermentation.[11] Like van Helmont, Sylvius frequently employed the process of fermentation, which required water, a mild heat and air,[12] although the last was not a necessity for van Helmont.

---

[8]Sylvius, *Disputationum medicorum decas* (1670) printed in *Opera medica* (1680), pp. 11–51 [Disputationum X, sects. 3, 4; p. 47).

[9]*Ibid., Disputationum* IV, sec. 1, p. 18.

[10]Nedham, *op. cit.*, p. 504; Sylvius, *op. cit., Disputationum* II, sec. 9, p. 13.

[11]Franciscus de la Böe, Sylvius, *A New Idea of the Practice of Physic; . . . The First Book; Of the diseases either Constitutions, producing, or following the Natural Functions of Man not in Health. Wherein is containd, beside a new Method in general, a Vindication of the Spleen and Mother from Fits attributed to them. As also a new Dioscovery of Intermitting Fevers, the Yellow Jaundice, and other Diseases never before discoverd. All cleard by Anatomical Experiments, and Chymical Demonstrations; As also by their Cures. Whereto is prefixed a Preface written Dr. Mar. Nedham. Trans. Richard Gower* (London: Brabazon Aylmer, 1675), pp. 4, 8, 28, 118. The work is a translation by an English student of Sylvius of the first book of the *Praxeos Medicae Idea Nova.*

[12]Sylvius, *op. cit., Disputationum* I, sec. 15, p. 12.

Digestion may be the process most frequently described by Sylvius to show the usefulness of chemistry as an explanatory model.[13] In this case the organs are treated as chemical laboratories and the complete process is seen as a series of chemical reactions, with an emphasis on three fluids, the saliva, pancreatic juice, and bile. Although he considered the first stage of digestion to be in the stomach, it was the saliva containing some salt and spirit that controlled this fermentation. Here a separation of the parts of the food resulted in the chyle. But this was only the first of a series of fermentations (separations) and effervescences (coagulations) in the intestines and heart that eventually resulted in the formation of the blood.

There is no need to describe this process in detail here since this has been done already by Lester King, but there are two points of special interest. First, in the course of his discussion Sylvius described the pancreatic fluid as mildly acidic and the bile as alkaline.[14] In theory the two could react through effervescence. Although experience did not indicate that this actually happened when the two were mixed, Sylvius held to his theory, which became one more basis for the developing acid-alkali theory. Second, Sylvius described the final production of blood in the heart in two stages. A first circulation occurred in the right chambers where it was partially perfected. Then it moved to the lungs for a completion of the process. Here the inspiration of air served to moderate the vital heat of the body, which was controlled by the heart. A salt in the air served as a cooling agent to maintain the proper heat and here he relied on the Paracelsian tradition of an aerial niter as the purest form of salt required for this final stage. As King has stated, "he concluded that in the lungs the inspired air, by virtue of its pure salt, serves to condense the blood previously rarefied in the right ventricle, and also to check the increased heat that has been aroused."[15]

Fever was also explained in chemical terms.[16] The acidic pancreatic juice might travel to the right ventricle of the heart, where an intense reaction with the chyle in the blood liberated fire particles and increased the pulse rate.

---

[13]Here see Lester S. King, *op. cit.*, pp. 98–107 and Audrey B. Davis, *Circulation Physiology and Medical Chemistry in England 1650–1680* (Lawrence, Kansas: Coronado Press, 1973), pp. 74–81. See also Partington, *op. cit.*, p. 284.

[14]*Ibid.*, pp. 284–85.

[15]King, *op. cit.*, p. 105. On the aerial niter as a Paracelsian concept, see Allen G. Debus, "The Paracelsian Aerial Niter," *Isis*, *55* (1964), 43–61.

[16]King. *op. cit.*, pp. 107–12.

Similarly, disease was attributed to an excess of either acid or alkali, a condition that could be overcome by neutralization. In this there is a reflection of the traditional medical cure by contraries rather than *similia similibus.*[17] There is certainly no doubt that Sylvius used and approved of chemically prepared remedies.

> Hitherto we have propos'd Medicines commonly known and approv'd of all; and if any being taught to be wiser than common People erect his Mind to *Chymical Preparations*, and Medicins more effectual, and so more safe, if they be both rightly prepar'd, and administered, we will commend unto him both *Tinctures*, and *Extracts*, and *Oils* artificially made of the foresaid, and *Volatil*, especially *Oily Salts* to be prepar'd by Art of most of the Parts of Living Creatures, and to be conveniently us'd: Which seeing they are usually not so ungrateful as common Medicins, may be taken in a much less quantity, and will work both sooner, and more mildly, and more effectually, and are deservedly desir'd of the Sick enough afflicted themselves: . . .[18]

And specifically,

> Among *Vomits*, I here prefer *Antimonials* before all others, both because they do most happily empty any Humors promiscuously, and because they are most friendly to Mans Nature, by degrees bringing all the Humors in Man after a peculiar manner, to a most laudable state; which power I doubt whether it be in other things likewise, but know it is in Antimonie; whatsoever Man blinded with their Prejudices prate to the contrary, both otherwise Learned and Unlearned.[19]

Sylvius also believed in the value of anatomical studies, which he sought to combine with his enthusiasm for chemistry. Here he is most frequently noted for having discovered the glandular tubercles in the lungs and for his observations on the brain.[20] And obviously he had a special interest in deter-

---

[17]Partington, *op. cit.*, p. 287.

[18]Sylvius, *New Idea*, p. 327.

[19]*Ibid.*, p. 174.

[20]Lindeboom, *op. cit.*, pp. 222–23.

mining the paths by which fluids could be transported throughout the body. Sylvius therefore played an important part in uniting anatomical studies with chemical explanations. Although this may be considered an important achievement by some, it was to pose a problem for contemporary Helmontians who rejected the study of human anatomy as a waste of time and effort.

THOMAS WILLIS (1621–1675) AND HIS CHEMICAL PHYSIOLOGY

The initial "Disputations" of Sylvius had been presented between 1659 and 1661, then published in 1663. The first publication of Thomas Willis appeared in 1659 and included the important *De fermentatione*, in which he clearly indicated the importance of chemistry. By this time Willis was already a well-known figure among English natural philosophers. Born in 1621, he had taken his B.A. at Christ Church, Oxford in 1639 and he had followed this with an M.A. three years later.[21] During the Civil War, he had served with the royalists in the defense of Oxford and had then returned to his studies, which resulted in a B. Med. degree in 1646.

By the late 1640s Willis was conducting chemical experiments in his rooms at Christ Church and he was an early member of a group interested in the sciences that included John Wilkins, Seth Ward, Jonathan Goddard, and later, Robert Boyle, Christopher Wren, Thomas Millington, and Robert Hooke. In addition to his continuing interest in chemistry, Willis participated in anatomical dissections. For him chemistry and anatomy were inseparably united. Although the *De fermentatione* was completed as early as 1656, it was not published until three years later with the *De febribus* and the short *De urinis* as *Diatribae duae medico-philosophicae.*

[21]The standard treatment of Willis is Dr. Hansruedi Isler, *Thomas Willis 1621–1675. Doctor and Scientist* (New York/London: Hafner, 1968; 1st edition in German, 1965). This is an excellent study that can be relied upon for the author's understanding of the importance of chemistry in Willis's thought. See also Partington, *op. cit.*, pp. 304–11 and Davis, *op. cit.*, pp. 12–44. The article on Willis in the *Dictionary of Scientific Biography* is by Robert G. Frank, Jr., *14* (1976), pp. 404–9. Frank has also discussed the views of Willis on fermentation in *Harvey and the Oxford Physiologists: A Study of Scientific Ideas* (Berkeley/Los Angeles/London: University of California Press, 1980), pp. 165–69. Of special importance is Audrey B. Davis, *Circulation Physiology and Medical Chemistry in England 1650–1680* (see above, note 13) where she discusses Willis throughout and has a special description of his views on fermentation on pp. 81–90.

Portrait of Thomas Willis (1621–1675). Courtesy of the National Library of Medicine.

With the end of the Protectorate and the return of Charles II, Willis was rewarded for his loyalty by being appointed the Sedleian Professor of Natural Philosophy in 1660. In the same year he took his doctorate in medicine. In succeeding years Willis went on to publish a series of works that combined his interests in chemistry and anatomy. These were collected in a folio volume (1676; English translation by Samuel Pordage, 1681 and 1684) that appeared posthumously and went through numerous editions until 1720.[22] His final years were spent in London, where he had a private practice and where he took an active role in the activities of the Royal Society.

It is in the early *De fermentatione* that we see most vividly the impact of current chemical thought on Willis. At first he thought that the subject was limited only to the baking of bread or the brewing of beer, but after some study his opinions changed: "I perceived I had gotten a far more large Province: Because it plainly appeared, besides these of Art, very many Works of Nature, to be not only like, but of themselves the effects of Fermentation."[23] Indeed, Willis was to look on fermentation as a fundamental key to the understanding of nature.

First, however, it was necessary to choose a basis for his discussion. Here he considered the elements of the Aristotelians, the atomism of Democritus and Epicurus, and the principles of the chemists. The first were quickly disposed of since it was little more helpful "to say an House consists of Wood and Stone, as a Body of four Elements." And, like van Helmont, Willis totally rejected fire as an element: "Some have dreamed, that its Sphear is scituate under the Moon: but this was introduced only, for the making the Hypothesis of the four Elements: for such a thing seems consonant neither to sense, nor Reason."[24] Nevertheless, in the 1650s atomism was gaining ever more acceptance. Indeed, the atomic hypothesis avoided occult qualities and sympathetic action. For Willis it "deserves no light praise" and may be termed "mechanical." But "it rather supposes, then demonstrates its Principles, and teaches of what Figure those Elements of Bodies may be, not what they have been, and also induces Notions extremely subtil, and remote from sense, and which do not sufficiently Quadrate with the Phaenomena of Nature. . . ."[25]

---

[22]The works of Willis were translated into English by Samuel Pordage as *Dr. Willis's Practice of Physick, Being the whole Works of that Renowned and Famous Physician* . . . (London: T. Dring, C. Harper and J. Leigh, 1681; also 1684 with reset type). Here the individual works are paginated separately. I have also used the Latin *Opera Omnia* (Venice: Sumptibus Ruïnetti, & Storti, 1708).

[23]From the preface to the *De fermentatione*, *Practice of Physick* (1684), sig. B2^r.

[24]*Ibid.*, p. 30.

[25]*Ibid.*, p. 2.

Here the hestitation of Willis may be compared with that of the royal chemist, Nicholas Lefèvre (ca. 1610–1669), who argued that the truths of chemistry overshadowed the speculation of the Schoolmen. As an example he turned to the composition of a body. The academic would explain that it was difficult to be certain for

> to be a body, it ought to have quantity, and consequently to be divisible: that a body ought to be composed of things divisible and indivisible, that is to say, of points and parts: but it cannot be composed of points, for a point is indivisible, and without quantity, and consequently cannot communicate any quantity to the body, since it hath none in it self, so that the answer should have concluded the body to be composed of divisible parts. But against this also will be objected, If it be so, let us know, whether the minutest part of the body is divisible or no, if it be answered, Divisible, then it is instanced again, that it is not the minutest, since there is yet a place left for division: but if this minutest part be affirmed to be indivisible, then the answer falleth again into the former difficulty, since it returns to affirm it a point, and consequently without quantity; of which being deprived, it is impossible it should communicate the same to the body, since divisibility is an essential property to quantity.[26]

Chemistry rejected "such airy and notional arguments" and stayed close to what could be actually demonstrated in the laboratory. Lefèvre held to the three Paracelsian principles plus earth and phlegm, a typical three-principle, two-element system of the period.

Here Willis and Lefèvre thought alike. Willis noted that fire analysis resolved bodies to Spirit, Sulphur, Salt, Water, and Earth. Therefore, "because this Hypothesis determines Bodies into sensible parts, and cuts open things as it were to the life, it pleases us before the rest." And yet, these principles need not be considered truly elementary since "I mean by the name of Principles, not simple and wholly uncompounded Entities, but such kind of Substances only into which Physical things are resolved into parts, lastly sensible."[27]

And yet, the small particles of bodies were not to be wholly ignored. Willis was well aware of the color changes brought about by chemical indi-

[26]Nicholas Lefèvre, *A Complete Body of Chymistry*, trans. P. D. C., 2 parts (London: O. Pulleyn, 1670), Part 1, pp. 9–10.

[27]Willis, *De fermentatione*, *Practice of Physick*, p. 2.

cators, long a subject of interest to chemists and soon to be a subject of investigation for Robert Boyle. Explaining these changes, Willis turned to the "minute" particles contained within the pores of every liquor

> which as to their site and position, being diversley altered, by another Liquor infused, transmit variously the Rays of Light, many ways break or reflect them, and so make divers appearances of colours. For when the Rays of Light pass through almost in right Lines, they make a clear colour, like Spring-water; but if in their passage, they be a little broken, the Liquor grows yellowish; but being more refracted, they cause a red colour; if they are bowed back so as to be drained, or that they cannot shew themselves, a dark or black colour arises; but if they are again reflected, to the outmost Superficies of the Liquor, they create the image of Whiteness: after this manner we might variously Philosophise about other colours, and their appearances, the diversity of which, and sudden alterations in Liquids, depend chiefly on Precipitation: because, as the Particles contained in the Liquor, are driven sometimes more near by another infusion that they clasp themselves together: sometimes are ordered into other series of positions, the diverse representations of colours is made. For Liquor being impregnated with little Bodies or Atoms, or this Nature, most minutely broken, seems as an Army of Soldiers placed in their ranks, who now draw into close Order, now open their Files and Ranks, now turn to the left, now to the right hand, as is diversley shown in the exercising of Tackticks, of the Art Military . . .[28]

Willis was also aware that salts crystallized in characteristic form, evidence that some chemical mechanists at the time were using in support of an atomic system of matter.[29]

However, useful as the corpuscular philosophy might be as an explanatory device in specific cases, it did not compare with the practical nature of the system of the chemists. Here the principles were made evident through laboratory procedures and fermentation served as an agent for natural change. A definition of the latter was required: "Fermentation is an intestine motion of Particles, or the Principles of every Body, either tending to the Per-

---

[28] *Ibid.*, pp. 40–1.

[29] *Ibid.*, p. 42.

fection of the same Body, or because of its change into another."[30] Fermentative particles exist everywhere and result in most of the changes that we observe in this world. Thus they are dispersed throughout the bowels of the earth which serves as a "pregnant Womb."[31] Here the ultimate result in his vitalistic system was the growth of metals and minerals. If, however, these particles were exhaled upwards from the earth's surface, the result was to be seen in meteors, and "we have read also, of Men changed into Stones, yea a whole City to have been stiffened into a stony substance, by the Air, or by some Vapour, brought forth of the Earth."[32] This is surely related to the existence of stony concretions. At the same time Willis argued that the fermentative action of saline particles resulted in the peculiar properties of spa waters and subterranean fires, while other vapors resulting from these particles caused wind, rain and snow.[33]

Similarly, the fermentative process was related to the life of vegetables and animals. Indeed, "the first beginnings of Life proceed from the Spirit Fermenting in the Heart," which moves quickly by circulating through the body and then returning to the heart.[34] As for the seminal vessels and the genital parts, they "do so swell up with Fermentative Particles, that there is nothing more: here Spirit, Salt, and Sulphur being together compacted, and highly exalted, seem in the Seed to be reduced as it were into a most noble Elixir."[35]

This is not all, for "We are not only born and nourished by means of Ferments; but we also Dye: Every Disease acts its Tragedies by the strength of some Ferment."[36] Fevers are kindled by the immoderate boiling of the sulphureous and spiritous parts of the blood. Scurvy, dropsy, the Stone and leprosy result from a flux of the saline part of the blood, which in turn results in various types of coagulations.[37]

In short, disease is due to fermentation, which is also its cure. Purging, vomiting, and sweating cool the blood and humors. Then cardiacs, digestives, chalybeats, and other medicines may be employed to restore the vigor and fermentation of the blood, "no otherwise than Wines growing sour, or

[30] *Ibid.*, p. 8.

[31] *Ibid.*, (1681 edition), p. 10.

[32] *Ibid.*, (1684), p. 9.

[33] *Ibid.*

[34] *Ibid.*, p. 11.

[35] *Ibid.*, p. 13.

[36] *Ibid.*, p. 14.

[37] *Ibid.*

degenerating into a deadness or want of strength, are mixed with more rich Lees, whereby they may Purge or grow turgid anew."[38]

Truly, the study of fermentation seemed to present a new understanding of both man and nature.

> Having thus far wandered in the spacious field of Nature, we have beheld all things full of Fermentation; not only in the distinct Provinces of Minerals, Vegetables and Animals . . . but also the whole sublunary World, seems as if one and the same substance were planted, and very pregnant through the whole with Fermentative Particles; which in every Region and Corner of it, as little Emmits in a Mole-hill, are busied in perpetual motion and agitation; they fly about here and there; sometimes upwards, sometimes downward they are hurried, they variously meet one another, associate themselves, and again depart asunder; with a continual Vicissitude they enter into divers Marriages, and suffer Divorces, on which the beginnings, the death, and transmutations of things depend. These little Bodies do not only very much abound in the bosom of the Earth, or in the midst of the waters, but they are especially diffused [throughout] the whole Atmosphear of the air in thick heaps.[39]

Printed with the *De fermentatione* were the *De febribus* and the *De urinis.* The former tract sought a cause of fever in the makeup of the blood, which Willis thought could be separated into the five chemical principles by means of distillation.[40] The same point was made in the *De urinis,* although the proportions of the principles in the two tracts were different.[41] And yet, analysis was not enough, because it was a fermentation caused by alien matter affecting the normal constituents that resulted in the "inordinate motion of the blood" with heat, thirst, and other symptoms variously disturbing the natural economy. The fever subsided only when the alien matter was subdued or excreted.[42]

As one might expect, Willis rejected the Galenic humors, stating that three of them were waste products.[43] In their place he substituted blood, a

[38] *Ibid.*

[39] *Ibid.*

[40] *De febribus, Practice* (1684), p. 48.

[41] *De urinis, Practice* (1684), p. 1 (separate pagination).

[42] Isler, *op. cit.*, pp. 69–70.

[43] *Ibid.*, p. 73.

nervous humor, and a serous humor. Cures were to be based on a chemical alteration of the blood through traditional evacuations plus digestives and a change in climate. This was to be followed by attention to diet.

Willis is famous for his anatomical study of the brain (1664). However, even here the motion of the blood and chemical analogies are repeatedly in evidence. Earlier in the *De fermentatione* he had described the brain as an alembic:

> . . . the Brain with a Scull over it and the appending Nerves, represent the little Head or Glassie Alembic, with a Spunge laid upon it, as we use to do for the highly rectifying of the Spirit of Wine: for truly the Blood when Rarified by Heat, is carried from the Chimny of the Heart, to the Head, even as the Spirit of Wine boyling in the Cucurbit, and being resolved into Vapour, is elevated into the Alembic; where the Spunge covering all the opening of the Hole, only transmits or suffers to pass through the more penetrating and very subtil Spirits, and carries them to the snout of the Alembic: in the mean time, the more thick Particles are stayed, and hindred from passing. Not unlike this manner, the blood being restrained within by the Skull, and its *Menynges*, as by an Alembick, are drunk up by the spungy substance of the Brain, and there being made more noble or excellent, are derived into the Nerves, as so many snouts hanging to it.[44]

Although much of the *Treatise on the Anatomy of the Brain* is purely anatomical, Willis was concerned with the purpose of the vessels. Thus he stated that the blood is supplied to the brain from "distillatory Vessels" and "the business of extracting the animal Spirits is performed even as a Chymical Elixir; to wit, great care is taken in the beginning of that Operation, both that choice of matter may be had, and that only a due proportion of it be exposed to the distillation."

> The blood by this means, as it were a Chymical work prepared, is carried by the fourfold chariot of the Arteries to four distinct regions of the Head . . . The blood being carried through the narrow infoldings and divarications of the Vessels as it were . . . through the serpentine chanels of an Alembick is made extremely subtile . . .[45]

---

[44]Willis, *De fermentatione* (1684), pp. 12–13.

[45]Willis, *Treatise on Anatomy of the Brain* (1684), p. 72.

Willis's awareness of contemporary chemical thought may also be seen in *An Essay on the Pathology of the Brain and Nervous Stock: In Which Convulsive Diseases Are Treated of.* Here he described muscular motion "to be no otherwise, but that the Spirits so expansed or stretched forth, as it were fired, after the manner of gunpowder, to be exploded or thrown out."[46] Here he referred specifically to the action of particles of nitre and sulphur, a concept that had been repeated frequently since the end of the previous century by Paracelsians.[47]

The lengthy *Pharmaceutice Rationalis* was written at the end of his life and the second part was published posthumously. Here Willis described the action of a wide range of traditional and chemical medicines. His purpose was not to prepare another pharmacopoeia, but rather, to find how medicines act in the body.

> But since almost the whole business of this *Pharmaceutick Drame* is acted behind the Curtain, therefore the various congressions of *Particles*, *Fermentations*, *Impulses*, and other diversities of Motions, which performed within, lye hid from the Senses, are to be searched out by a more deep Scrutiny of the Intellect, which if it may be at last performed, there is nothing more will be desired, whereby *Physick* being performed in all its parts, may grow to a true Science and be practised with greater certainty, not inferiour to the Mathematicks.[48]

He severely criticized those pseudo-chemists who promised great elixirs, panaceas, and a universal medicine while working "without any Counsel or direction of Method, but by meer chance and with a certain blind understanding."[49] In contrast, Willis took as his object to find (1) how medicines affect animal spirits, and (2) how they ferment the blood and humors for "from thence a Path would lead to the unfolding the *Reasons* of every Evacuation."[50]

---

[46]Willis, *An Essay on the Pathology of the Brain and Nervous Stock: In Which Convulsive Diseases are Treated of* (1684), p. 2.

[47]Allen G. Debus, "The Paracelsian Aerial Niter," *Isis*, *55* (1964), pp. 43–61.

[48]Willis, *Pharmaceutice Rationalis* (1684), Part 1, Preface, sig. A2^{v}.

[49]*Ibid.*, Part 1, sig. A3^{v}.

[50]*Ibid.*, Part 1, sig. A4^{r}.

The *Pharmaceutice Rationalis* may be of most value for its description of diseases and here his long discussion of *diabetes mellitus* is frequently noted.[51] But Willis also gave his blessing to bloodletting, noting that "*Helmont* of late, and still certain followers of him, Pseudochymists and Fanaticks, have ejected Bleeding out of all Physick, because they think this evacuation to be a great injury to nature."[52] The *Pharmaceutice Rationalis* shows a close familiarity with chemical preparations and it occasionally preserved interesting accounts of seventeenth-century medical quackery such as the following:

> Mountebanks do ordinarily prescribe the smoak of Arsenick to be suckt into the mouth, like Tobacco kindled in a Pipe, and sometimes with good success: Moreover it is in practise with the Vulgar, to burn like Tobacco in a Pipe little bits of cloth stained with Arsenick (such as wherewith the walls of Taverns are hung) and so suck the smoak into the consumtive Lungs for cure.[53]

It is important to note that although both Willis and Sylvius were convinced of the value of chemical explanations, they differed on a number of points. Above all, Sylvius developed the acid-alkali concept while Willis emphasized fermentation far more. And if neither of them adopted the more mystical tenets of the Helmontian *opera*, both were convinced of the significance of the Harveian circulation.

ROBERT BOYLE (1627–1691) AS AN HELMONTIAN CHEMIST

It is customary for Robert Boyle to figure more prominently in the history of chemistry than in the history of medicine. At one time, based largely on his *Sceptical Chymist*, he was termed the "Father of Chemistry," a title that may well be questioned today. He is also normally classified among the mechanical philosophers as a champion of seventeenth-century atomism. However, during his lifetime the work of chemists was almost always associated with medicine. Boyle's work was to be no exception to this rule, as we shall see. The medical implications of his work were those of most concern to Marchamont Nedham and others.

---

[51] *Ibid.*, Part 1, p. 71.

[52] *Ibid.*, Part 2, p. 131.

[53] *Ibid.*, Part 2, p. 44.

Portrait of Robert Boyle (1627–1691). Courtesy of the National Library of Medicine.

Today the positivist view of Boyle is changing rapidly. This is due in part to the ongoing project involving the cataloging and study of his papers, but a view of Boyle unbiased by a presentist viewpoint began to emerge more than thirty years ago.[54] As we look at his earliest writings, we find a young man profoundly influenced by the work of van Helmont. The mechanist Boyle developed only later and this reaffirms the importance of approaching him and other scientists chronologically. It is interesting to note that Boyle gradually lost his enthusiasm for van Helmont, but there is nothing to indicate that he ever lost his belief in the possibility of alchemical transmutation.

---

[54]A bibliography of the recent work on Boyle appears in Michael Hunter, ed., *Robert Boyle Reconsidered* (Cambridge University Press, 1994). Three papers of special importance in this volume are those by Lawrence Principe, "Boyle's Alchemical Pursuits," pp. 91–106; William Newman, "Boyle's Debt to Corpuscular Alchemy," pp. 107–18. Here Newman clearly shows the impact of the medieval Geberian texts on Boyle's corpuscular thought. See also Antonio Clericuzio, "Carneades and the Chemists: a Study of the *Sceptical Chymist* and its Impact on Seventeenth-Century Chemistry," pp. 79–90. Among Clericuzio's other papers relevant to this section is his "A Redefinition of Boyle's Chemistry and Corpuscular Philosophy," *Annals of Science, 47* (1990), 561–69. Professor Hunter is in the process of preparing a new edition of Boyle's works and his papers. A paper by him of interest is "Alchemy, magic and moralism in the thought of Robert Boyle," *British Journal of the History of Science, 23* (1990), 387–410. Partington's account, *op. cit.*, 2, pp. 486–549 remains authoritative for Boyle's chemistry. An essential work on Boyle's alchemy is Lawrence M. Principe, *The Aspiring Adept: Robert Boyle and his Alchemical Quest* (Princeton: Princeton University Press, 1998). In *The Diffident Naturalist: Robert Boyle and the Philosophy of Experiment* (Chicago: The University of Chicago Press, 1995), Rose-Mary Sargent agrees that "Of all the chemists, Boyle liked the Helmontians best, [but even] they did not escape his criticism . . . ," p. 72. Barbara Beigun Kaplan discussed aspects of Boyle's medical views that went beyond his medico-chemical views in *Divulging of Useful Truths in Physick: the Medical Agenda of Robert Boyle* (Baltimore: Johns Hopkins Press, 1993). Although noting (p. 35) that Boyle sought to unite iatrochemical with atomic views, she believes that Boyle was less interested in "articulating the idea of the body as a perfectly mechanical self-contained system than he was in illustrating how the living body reacted with the external world and how both were ultimately part of a single dynamic system designed by God." (p. 75)

Marie Boas's *Robert Boyle and Seventeenth-Century Chemistry* (Cambridge University Press, 1958) attempted to present Boyle as a major figure of the late seventeenth century for reasons other than those associated with the *The Sceptical Chymist.* My criticism of her work appeared in "The Significance of the History of Early Chemistry," *Cahiers d'Histoire Mondiale, 9* (1965), 39–58. I later showed the dependence of Boyle on van Helmont in "Fire analysis and the Elements in the Sixteenth and the Seventeenth Centuries," *Annals of Science, 23* (1967), 127–47. In *The Chemical Philosophy* (1977), I expanded this work to show the Helmontian nature of Boyle's early *Usefulness of Experimental Philosophy, 2*, pp. 473–92.

Although the new edition of the works of Boyle is currently ongoing [the first seven volumes as *The Works of Robert Boyle*, ed. Michael Hunter and Edward B. Davis (London: Pickering and Chatto, 1999), see the review by Peter Anstey in *On the Boyle*, no. 3 (December, 1999)], the present chapter was completed earlier and most references will be made to the 1772 edition (see below, note 56).

Here, however, we will be concerned primarily with some of his early writings that were of concern to contemporary physicians.[55]

Robert Boyle was born in 1627 with every advantage. He was the fourteenth child of the Earl of Cork and educated at Eton. He left for a tour of the Continent in 1638 with the guidance of a tutor and did not return home until 1644. Ten years later he moved to Oxford, where he became an active member of the group that included Thomas Willis. During this period he employed Robert Hooke as his assistant and was instrumental in bringing Peter Sthael (or Stahl) to Oxford to teach a course in practical chemistry. Boyle also worked with the Harvard graduate, George Starkey (d. 1665), who dedicated his *Pyrotechny Asserted and Illustrated* (1658) to him. It was not until 1668 that he moved permanently to London, where he lived with his sister, Lady Ranelagh.

Boyle's early interest in chemistry may be seen in a letter of March 6, 1646/7, in which he lamented the breaking of an earthenware furnace while enroute to him on the rough roads. Although the chemical glassware had survived, it was of no use without the furnace.[56] A month later (8 April 1647) he wrote to Samuel Hartlib that he hoped to see translations made of Tommaso Campanella's *Civitas solis* and Johann Valentine Andreae's *Christianopolis*,[57] both "scientific" utopias of interest to contemporary chemists. And in a letter to his brother, Lord Broghill (19 June 1652), he wrote that "those excellent sciences, the mathematics," were "the first I addicted myself to, and was fond of," but that "experimental philosophy with its key, chemistry," had succeeded "them in my esteem and applications."[58] This shift of interest from mathematics to chemistry reflects the new interest in chemistry and chemical medicine that followed the publication of van Helmont's works in 1648.

Boyle was always interested in religion. His *Style of the Scriptures* appeared in 1661, but it was an essay that he had been working on for some years.[59] This

---

[55]Although only some of his early writings will be referred to here, it is important to note that Boyle was concerned with medicine throughout his life. See Michael Hunter, "Boyle Versus the Galenists: An Aborted Revolution in Seventeenth-Century Medicine" in a paper which the author kindly permitted me to read in typescript. Here he refers to a work critical of traditional medicine that was never published.

[56]Thomas Birch, "The Life of the Honourable Robert Boyle" in *The Works of the Honourable Robert Boyle* (6 vols., London: J. and F. Rivington et al., 1772), *1*, pp. xxxvi–xxxvii.

[57]*Ibid.*, *1*, pp. xxxviii.

[58]*Ibid.*, *1*, p. xlix.

[59]John F. Fulton, *A Bibliography of The Honourable Robert Boyle, Fellow of the Royal Society* (2nd ed., Oxford: Clarendon Press, 1961), p. 33.

combined interest in chemistry and religion may be frequently found in the works of Paracelsians and Helmontians. We are then hardly surprised to see it in Boyle's *Of the Usefulness of Experimental Philosophy principally as it relates to the Mind of Man.* Although the first part of this work with much of the second part appeared in 1663, Boyle informed the reader that a great portion of the first part had been written when he was "scarce above 21 or 22 years old"[60] (1648 or 1649). The second part of the work was surely later, but written prior to 1658.

In part one we find a typical Paracelsian view of man being urged to study nature as a religious act. It is far more difficult to investigate nature through the experimental philosophy than to learn the works of the heathen Greeks through learning by rote at the schools and colleges.[61] The Aristotelians explained nature without God. But we as Christians may turn to the first chapter of Genesis for a knowledge of the Creation. Through a study of God's work we honor and praise Him: "I esteem the world a temple . . . it will not be rash to infer, that if the world be a temple, man sure must be the priest, ordained . . . to celebrate divine service, not only in it, but for it."[62] Even more telling is Boyle's citation from the *De mineralibus* of Paracelsus: "God is very admirable in his works: from the Contemplation of which we ought not to desist night or day, but continually be imployed in the inquisition of them: *for this is* to walk in the ways of God."[63] At this time Boyle may still have questioned the existence of atoms, but he was convinced that the Helmontian seminal principle was required for an explanation of the regular formation of crystals and the growth of minerals. These instances clearly negated the atheistic viewpoint of the Aristotelians.[64]

The first section of the second part *Of the Usefulness of Experimental Philosophy* was devoted to its usefulness to medicine. Here we are told that experimental philosophy will not only give us a greater knowledge of our Creator, but will also lead to a greater mastery over nature. This, in turn, should be of special value to the physician. Following van Helmont here, Boyle rejected the views of those Paracelsians who believed that man was a perfect microcosm of the great world. And yet, again like van Helmont,

[60]Robert Boyle, *Some Considerations Touching the Usefulness of Experimental Philosophy*, in *Works 2*, Part 1, p. 4.

[61]*Ibid.*, p. 5.

[62]*Ibid.*, p. 32. See also Harold Fisch, "The Scientist as Priest: A Note on Robert Boyle's Natural Theology," *Isis*, *44* (1953), 252–65.

[63]Boyle, *Usefulness*, p. 57.

[64]Boyle's critique of the atomic philosophy will be found in *ibid.*, pp. 40–49.

> certain is it, that there are many productions, operations, and changes of things, which being as well to be met with in the great, as in the little world, and divers of them disclosing their natures more discernibly in the former, than in the latter; the knowledge of the nature of those things, as they are discoverable out of man's body, may well be supposed capable of illustrating many things in man's body, which receiving some modifications there from the nature of the subject they belong to, pass under the notion of the causes or the symptoms of diseases.[65]

Traditional medicine was founded on humoral theory. Here Boyle commented that

> though I am very unwilling to meddle with medical controversies, and am apt to think that chymists are wont to speak somewhat slightingly of the humours of the human body, and allow them too little a share in the production of diseases; yet [I am inclined] to believe, that the generality of former physicians have ascribed too much to the humours, under the notion of their being hot and dry, cold and moist, or endowed with such other elementary qualities, and have taken too little notice of the saline (if I may so speak) and suphureous properties of things.[66]

For Boyle, chemistry held great promise for medicine. He disapproved of the "vain-glorious boasts of Paracelsus," but he felt that the chemists were owed something "for having put some men in hope of doing greater cures than have been formerly aspired to, or even thought possible . . ."[67]

> it may be safely affirmed, that not only physiology, in its full extent, but that handmaid to it, which is called chymistry, may not a little contribute to clear up the nature of both digestions, and of those deficiencies or aberrations in them, which produce a great part of diseases . . .[68]

To be sure, chemistry will not explain all pathological phenomena, but "true chymistry" will help the physician to understand many of them "which can scarce be solidly explicated without it."

[65] *Ibid.*, Part 2, Sect. l, p. 76.

[66] *Ibid.*, p. 80.

[67] *Ibid.*, p. 90.

[68] *Ibid.*, p. 79.

> And let me add, that he, that thoroughly understands the nature of ferments and fermentations, shall probably be much better able than he, that ignores them, to give a fair account of divers phaenomena of several diseases (as well fevers as others).[69]

Boyle was critical of those chemists who ignored anatomy and here he surely reflected the union of chemistry and anatomy being effected by both Willis and Sylvius.[70] In addition, he had nothing but praise for the chemical search for new remedies. These were far superior to the traditional *materia medica.*

> Some Paracelsian would, perhaps, set forth, how much more easy to be taken chymically prepared medicines are wont to be, than those loathesome and clogging Galenical potions, bolus's, &c. which are not only odious to the takers, but (which is much worse) are to many so offensive, that either the patients cannot get them down, or the incensed stomach returns them by vomit, before they have stayed long enough in the body to do any more than distemper it.[71]

Not only were these chemicals easier for the patient to take, they were also more powerful and cheaper because they were used in smaller amounts than the costly Galenic preparations.[72]

Chemistry then would prove valuable for the preparation of new remedies as well as for the examination of diseases that had been caused by fermentations. It should also be of value in the study of bodily fluids and juices. Boyle was well aware of the current interest in the infusion of chemical remedies into the blood stream. Because of the general acceptance of the circulation of the blood this seemed to be a promising way to direct medicines to all parts of the body.[73]

---

[69] *Ibid.*, p. 81.

[70] *Ibid.*, p. 83. Here Boyle refers to the need for vivisection.

[71] *Ibid.*, p. 114.

[72] *Ibid.*, p. 132.

[73] *Ibid.*, pp. 88–89. For a description of the early transfusion and infusion experiments, see Davis, *Circulation Physiology*, pp. 178–83. Michael T. Walton and Phyllis J. Walton discuss pre-Harveyan references to the significance of the blood and give the transfusion texts in Jerome Cardan's *De rerum Varietate* (1558) and Andreas Libavius's *Appendix necessaria syntagmatis arcanorum chymicorum* (1615) in "Witches, Jews, and Spagyrists: Blood Remedies and Blodd Transfusion in the Sixteenth Century," *Cauda Pavonis, 15* (1996), 12–15.

Throughout all of this we are constantly referred to van Helmont. Indeed, Boyle stated that "I must here confess to you once and for all, that (always excepting his extravagant piece, *De magnetica vulnerum curatione*) I have not seen cause to disregard many things he delivers, as matters of fact, provided they be rightly understood; . . ."[74] And yet, although he rejected the work on the magnetic cure of wounds, Boyle did not hesitate to describe remarkable subjects and cures himself. In regard to van Helmont's much vaunted universal solvent, the alkahest, Boyle hesitated to believe in its existence "till sufficient experience hath convinced me of [its] truth."[75] But then, he added, perhaps special solvents could exist and could be directed at bladder stones, which were the bane of the upper classes. After all, he observed, the magnet operates on iron alone. Why not a solvent that operated only on stones formed in the body?[76]

He was interested also in the medicinal properties of moss from human skulls, which some chemical authorities affirmed to have a stanching property on blood.[77] Medical transplantation was also widely discussed in the fifties. In this case it was thought that a disease could be transplanted from a person to a plant or an animal. Boyle remarks that Robert Fludd "transplanted the gout of one of his patients by making him often sleep with a dog, that was fond of him, who thereby became subject to such periodical fits of the gout as the master had been troubled with."[78] Similarly,

> Dr. [William] Harvey, who, as rigid a naturalist as he is, scrupled not often to try the experiment mentioned by *Helmont*, of curing some tumours or excrescencies, by holding on them for a pretty while (that the cold might thoroughly penetrate), the hand of a man dead of a lingering disease, which experiment the doctor was long since pleased to tell me, he had sometimes tried fruitlessly, but often with good success.[79]

When it came to sympathetic medicine, Boyle clearly was a sceptic, but he admitted that it did seem to work sometimes. That was enough to classify

---

[74]Boyle, *Usefulness*, p. 149.

[75]*Ibid.*, p. 97.

[76]*Ibid.*

[77]*Ibid.*, p. 177.

[78]*Ibid.*, p. 166.

[79]*Ibid.*

such remedies as medicines "especially since these sympathetical ways of cure are most of them so safe and innocent, that though, if they be real, they may do much good, if they prove fictions they can do no harm . . ."[80]

The final part of *The Usefulness of Experimental Philosophy* (actually the remainder of the second section) appeared in 1671 and was composed of six essays. Here experimental philosophy was related to man's empire over inferior creatures, mathematics, goods and trades, physical knowledge, and man's ignorance of the use of natural things. We are told that the work was written about 1658[81] so that once again we are dealing with a composition completed long before its actual publication.

The essay on the value of mathematics here is of special interest in a period when increased interest in quantification was influencing the development of the sciences. In a 1652 letter to his brother, Boyle had admitted that his early interest in mathematics had been overtaken by a new interest in chemistry. Now, six years later, he advised Pyrophilus not to "be influenced, by the great auhority of those modern philosophers, who would have the use of mathematicks, as disciplines, that consider only abstracted quantity and figure, to be rather hurtful than advantageous to a naturalist, the object of whose studies ought to be matter." Rather,

> I think that a competent knowledge in mathematicks (for a profound one is not always necessary) may be so serviceable to those that would become philosophers, that I shall not scruple to mention it as another thing, which may increase your expectation from physiology, that those, who pass for naturalists, have, for the most part, been very little, or not at all, versed in the mathematicks, if not also jealous of them.[82]

This passage is of interest because although Boyle maintained his primary interest in chemistry, his views were changing from an original Helmontian bias to a view more closely oriented toward a mathematically based mechanical philosophy.

One additional early work by Boyle must be mentioned here, *The Sceptical Chymist.* Published first in 1661, this is another case of a work that under-

---

[80] *Ibid.*, p. 168.

[81] Boyle, *Some Considerations Touching the Usefulness of Experimental Natural Philosophy . . . The Second Tome, containing the latter Section of the Second Part* in *Works, 3*, p. 395.

[82] *Ibid.*, p. 426.

went a long gestation. An early version of the *Sceptical Chymist*, titled "Reflexions on the experiments vulgarly alledged to evince the 4 peripatetique elements, or ye 3 chymicall principles of mixt bodies." was published by Marie Boas, who dated the work to the early 1650s.[83] This date has been questioned by others, but Clericuzio agrees with Boas in the most recent discussion of the piece.[84] This manuscript version clearly shows a strong Helmontian influence, as we would expect from this period. Nor need we be surprised to find that the published work still reflects Boyle's early enthusiasm for the Belgian authority.

Boyle began by noting that chemistry was beginning to attract far more adherents than in the past.[85] However, he was concerned about the state of element theory, which he now proposed to examine. Early in the text he turned to atomism as a basis for his own natural philosophy. In a series of propositions he began by stating that it seemed reasonable "to conceive that at the first Production of mixt Bodies, the Universal matter whereof they among other Parts of the Universe consisted, was actually divided into little Particles of several sizes and shapes variously mov'd."[86] These then may be gathered together into primary concretions and more complex forms of matter.

In criticizing the elemental system of the Aristotelians as well as the *tria prima* of Paracelsus, Boyle hoped to bring forth abler chemists, who "will be oblig'd to speak plainer than hitherto has been done, and maintain [their system] by better Experiments and Arguments . . ."[87] And yet, Boyle did not reject those chemists who founded their beliefs on the Mosaic account in Genesis.[88] Nor did he believe that there need be conflict between atomism and the Mosaic philosophy.

There is no doubt that Boyle was a lifelong believer in the possibility of transmutation. His tract on the degradation of gold (1678)[89] and his move for the repeal of Henry IV's Statute (1404) against the multiplication of gold

---

[83]Marie Boas, "An Early Version of Boyle's *Sceptical Chymist*," *Isis*, *45* (1954), 153–68.

[84]Clericuzio, "Carneades and the Chemists" in Hunter, *Boyle Reconsidered*, p. 80.

[85]Robert Boyle, *The Sceptical Chymist or Chymico-Physical Doubts and Paradoxes, Touching the Spagyrist's Principles Commonly Call'd Hypostatical: As they are wont to be Propos'd and Defended by the Generality of Alchymists. Whereunto is Praemis'd Part of another Discourse relating to the same Subject* (London: J. Cadwell for J. Crooke, 1661; London: Dawson Reprint, 1965), sig. A2^{r}.

[86]*Ibid.*, p. 37.

[87]*Ibid.*, sig. A5^{v}.

[88]*Ibid.*, p. 38.

[89]*Of a Degradation of Gold Made by an Anti-Elixir: A Strange Chymical Narrative* (1678).

(1689)[90] are well known. Currently Lawrence Principe's laboratory work on Boyle's processes and William Newman's investigation of Boyle's papers have done much to flesh out our knowledge of Boyle's alchemical interests.[91] In the *Sceptical Chymist*, we find evidence of this obsession. Here he distinguishes between the chemical "cheats" and the true adepts, "by whom, could I enjoy their Conversation, I would both willingly and thankfully be instructed; especially concerning the Nature and Generation of Metals."[92] Even more interesting is the fact that although he constantly attacked the mystical and obscure language of many alchemists, he thought it quite proper to "excuse the Chymists when they write Darkly, and Aenigmatically, about the Preparation of their *Elixir*, and some other grand *Arcana*, the divulging of which they may upon grounds Plausible enough esteem unfit."[93]

The chemical texts of the period invariably began with a discussion of the elements and for this reason Boyle centered his discussion on them. He dismissed the four elements quickly, arguing that the flame, the smoke, the sap, and the ash from a burning green twig did not prove that it was made of fire, air, water, and earth.[94] In fact, he argued, fire does not separate substances into their elements. Rather, it alters them by making new combinations.

In turning to the chemical elements, Boyle took up the question of the validity of fire analysis in detail, showing that not all substances were affected by fire and that of those that were, not all resulted in the same number of fractions. Furthermore, far different results were obtained from fire and distillation. Here Boyle borrowed heavily from van Helmont's discussion of the same subject without referring to the origin of the argument, a topic I have examined earlier in detail.[95] He found van Helmont's willow tree experiment a possible solution to the problem since this indicated that vegetable

---

[90]Birch, *Life* in Boyle, *Works*, *1*, pp. 131–32.

[91]See particularly the papers referred to above in note 54. See also Lawrence M. Principe, "The Gold Process: Directions in the study of Robert Boyle's Alchemy" in Z. R. W. M. von Martels, ed., *Alchemy Revisited: Proceedings of the International Conference on the History of Alchemy at the University of Groningen 17–19 April 1989* (Leiden et al.: E. J. Brill, 1990), pp. 200–6, Principe, "Robert Boyle's *Dialogue on Transmutation*," a paper presented at the History of Science Society Meeting in Madison, Wisconsin, November, 1991 and Principe, *Aspiring Adept, passim.*

[92]*Sceptical Chymist*, sig. A7^r.

[93]*Ibid.*, p. 203.

[94]*Ibid.*, pp. 27–29.

[95]*Ibid.*, pp. 49–51. See also Debus, "Fire Analysis and the Elements in the Sixteenth and Seventeenth Centuries" (above note 53).

matter is composed largely of water. Boyle grew plants in water, which he then distilled, and it seem evident

> that water may be transmuted into all the other elements: from whence it may be inferred, both, that 'tis not everything chymists will call salt, sulphur, or spirit, that needs alwaies be a primordiate and ingenerable body. And, that nature may contex a plant (though that be a perfectly mixt concrete) without having all the elements previously presented to her to compound it of.[96]

In short, the three principles themselves may well be made of water. Boyle was by no means fully convinced of a primal water, but his high regard for van Helmont made this concept one that required serious consideration.

Whether or not there are three principles, or more, the ones proposed by the chemists do not explain all natural phenomena and they should not be considered the proper basis for natural philosophy. Boyle was indeed a "sceptical" chemist. He felt that perhaps either system "may be much more probably maintained than hitherto it seems to have been . . ."[97] However, fire analysis was not a satisfactory method of analysis and whatever their beliefs, chemists would have to devise better experiments to support their views.

COOPERATION AND CONFRONTATION: THE HELMONTIAN MECHANISM OF THOMAS SHERLEY AND THE SOCIETY OF CHEMICAL PHYSICIANS

In the short summary we have given of the chemical and chemically related views of Sylvius, Willis, and Boyle, we have held closely to writings of the late 1640s and the decades of the fifties and the early sixties. In all cases we have seen an intense interest in the relation of chemistry to medicine. Both Sylvius and Willis combined their chemical interests with anatomy and Sylvius developed in more detail than van Helmont the interaction of acids and alkalis. Willis was more interested in fermentation, but both men were convinced of the value of chemistry for medicine. Boyle, although not a physician himself, was also convinced of the importance of chemistry for medicine. His

---

[96] *Ibid.*, p. 352.

[97] *Ibid.*, pp. 435–36.

early *Usefulness of Experimental Philosophy* reflects much of the Paracelsian and Helmontian traditions in regard to both natural philosophy and medicine. We have here three major figures who reflected the current interest in chemistry and medicine, all of whom may in some fashion show the influence of the Paracelsian tradition either through the work of van Helmont or earlier authors.

There is little doubt that chemistry was widely discussed in the mid-seventeenth century. This has been well documented for the English scene by Antonio Clericuzio.[98] In England, Willis replied to Boyle's criticism of the principles by adding a defense of their use in the third edition of his *De fermentatione.*[99] Other chemists sought to establish a new philosophy based on a union of Helmontian and mechanical principles. Thus, Thomas Sherley (1638–1678) wrote *A Philosophical Essay . . . From which occasion is taken to search into the Origin of all Bodies, discovering them to proceed from Water, and Seeds* (1672).[100] Here Sherley sought to define a new philosophy that would contribute to the advance of medicine as well as to a more general understanding of nature. He found his answer in a union of the mechanical philosophy and the chemical philosophies. For the first he turned to Robert Boyle's *Usefulness of Experimental Philosophy* and *The Sceptical Chymist.* For the second he turned to van Helmont's concept of water and seed.[101]

Above all, Sherley believed that van Helmont had introduced to natural philosophy a certain explanation of motion. This was fundamental since motion was basic for the mechanical philosophy. And what was this founded on?–water and seed. In the course of his discussion Sherley rejected both the Aristotelian elements and the principles of the chemists. He followed both Boyle and van Helmont in their critique of fire analysis[102] and he used this as the basis of his attack on Thomas Willis's defense of the five chemical principles.[103] Sherley devoted considerable space to van Helmont's alkahest

---

[98]Antonio Clericuzio, "From van Helmont to Boyle, A Study of the Transmission of Helmontian Chemical and Medical Theories in Seventeenth Century England," *British Journal of the History of Science, 26* (1993), 303–34.

[99]Noted by Clericuzio in "Carneades and the Chemists" in Hunter, *Robert Boyle Reconsidered,* p. 85.

[100]See Allen G. Debus, "Thomas Sherley's 'Philosophical Essay' (1672)," *Ambix, 27* (1980), pp. 124–35.

[101]Thomas Sherley, *A Philosophical Essay: Declaring the Probable Causes, whence Stones are produced in the Greater World. From which occasion is taken to search into the Origin of all Bodies, discovering them to proceed from Water, and Seeds* (London: William Cademan, 1672), sig. A6^v.

[102]*Ibid.,* p. 84.

[103]*Ibid.,* pp. 85–97.

because its ability to reduce all substances to water gave great weight to the belief that the primal substance was water.[104]

Sherley had prepared a book calling for a new philosophy, one that would unite the work of the chemists and the mechanists. He was able to do this by concentrating on the early works of Boyle that were surely inspired by van Helmont. Like both of them, Sherley was convinced that one must begin with a proper understanding of the true elements of nature. He correctly saw the connection between van Helmont's critique of the Aristotelian elements on the one hand, and Boyle's *Sceptical Chymist* on the other. Indeed, the *Philosophical Essay* may best be characterized as an avowedly Helmontian revision of *The Sceptical Chymist.*

Surely more indicative of the determination of the Helmontian physicians was their attempt to found a medical society of their own, the Society of Chymical Physicians, to rival the Galenist Royal College of Physicians. The evidence for this society was first presented in the now classic paper by Sir Henry Thomas in 1953 as the result of the discovery of a broadside licensed 28 June 1665.[105] Although members of the nobility and the Archbishop of Canterbury supported the chemists, their numbers were decimated during the Plague of London (1665) while tending to the sick. A year later the Great Fire of London raised new weighty problems. In the end nothing came of their grand plan for a new medical society.[106]

The June 28th "Advertisement" from the Society of Chemical Physicians was signed by William Goddard, Marchamont Nedham, Edward Bolnest, Thomas Williams, Everard Manwaring, George Thompson (Thomson), Thomas O'Dowde, and Richard Barker. Another list dating from 1665 by O'Dowde offers thirty-five names, including those of the Lord Chancellor

---

[104] *Ibid.*, pp. 41–43.

[105] Sir Henry Thomas, "The Society of Chymical Physicians: an Echo of the Great Plague of London, 1665" in E. Ashworth Underwood, *Science, Medicine and History: Essays on the Evolution of Scientific Thought and Medical Practice written in honour of Charles Singer* (2 vols., Oxford University Press, 1953), *2*, pp. 56–71.

[106] The fortunes of the Chymical Society have been studied by a number of scholars since the Thomas article appeared. Here I refer particularly to P. M. Rattansi, "The Helmontian-Galenist Controversy in Restoration England," *Ambix, 12* (1964), 1–23; Harold J. Cook, "The Society of Chemical Physicians. The New Philosophy, and the Restoration Court," *Bulletin of the History of Medicine, 61* (1987), 61–77; and the Clericuzio articles referred to earlier, especially "From van Helmont to Boyle" (note 98). An earlier reference that is still of interest is R. F. Jones, *Ancients and Moderns: A Study of the Rise of the Scientific Movement in Seventeenth-Century England* (1936; 2nd ed., St. Louis: Washington University Press, 1963), pp. 206–16.

and the Duke of Buckingham.[107] Although several of these chemists presented their views in print, here we are most interested in Marchamont Nedham, the author of the *Medela medicinae.*

MARCHAMONT NEDHAM AND THE *MEDELA MEDICINAE*

Marchamont Nedham (1620–1678) was educated at Oxford, where he was a student at All Souls' College and took his B.A. degree in 1637.[108] Although he had studied medicine, he had a talent for journalism. He published the *Mercurius Britannicus* beginning in August, 1643. At that time he was an enemy of the royal cause and was arrested. After being released, he turned to medicine. But in 1647 we find him again publishing, this time in favor of the King, in a royalist newspaper, the *Mercurius Pragmaticus.* Arrested again, he was released and proceeded to publish a pro-Commonwealth paper, the *Mercurius Politicus* (1650). With the royalists now demanding his death, Nedham fled to the Netherlands when the Restoration of the King occurred in 1660. However, he was pardoned by Charles and turned once more to medicine.

Nedham's *Medela Medicinae. A Plea for the Free Profession; and Renovation of the Art of Physick . . . Tending to the Rescue of Mankind from the Tyranny of Diseases; and of Physicians themselves from the Pedantisn of old* Authors *and present Dictators* was completed in late 1664. The license is dated 2 September and Nedham's "Epistle didicatory" to the Marquis of Dorchester is dated 26 November. Published early the following year, this book is clearly a call for a new chemical medicine to replace the tradition-bound Galenism of the Royal College of Physicians. No matter that the situation was not as bleak as it was pictured by Nedham–he painted a disturbing picture of the medical establishment, and did so in the journalistic style that makes his book a rapid read compared to many others of the period.

Nedham deplored the continued reliance on ancient authorities in medicine. Because of this, one could see a real difference between the state of medicine and the other sciences.

> . . . *the Liberal Sciences have fared far worse than the Arts Mechanical, which always have been, and are continually much improved*; whereas in

[107]Cook, *op. cit.*, pp. 70–71.

[108]The detailed biographical note in the *Dictionary of National Biography* is by C. H. Firth [21 vols., Oxford University Press, 1920–1921, *14*, pp. 159–64.

> this of Physick and Philosophy, men have formerly thought it wisdom not to budg an inch from the footing of the first masters; insomuch that when *Chymistry* first came in play, the Professors and Operators were thought to be Mad-men . . .[109]

And worse, the Aristotelians and Galenists have persecuted the chemists. Van Helmont had been "reviled as an enthusiast and a Mad-man,"[110] while even William Harvey had been attacked when he had first published his *De motu cordis.*[111]

Nedham willingly admitted that "a Doctor bred upon in the Contemplative Philosophy of the Schools, may be a Scholar and a very fine Gentleman; but what is that to the Curing of a Disease, or the rousing of a Heart-sick Man from his bed of Languishment?"[112] These gentlemen-physicians defend a science developed by "gross Heathens" and speak of a "*Divine Hippocrates, Divine Galen* and I know not how many more titles more fit for God than men . . ."[113]

In his call for medical change, Nedham placed great stress on the fact that diseases had altered with time. This is a point noted by earlier Paracelsians, but Nedham turned to statistical evidence to show the changing patterns of death in England. Here he referred to John Graunt's study of the Bills of Mortality for London. As an example, rickets had appeared as a cause of death for the first time in 1634.[114] Thirty years later it was far more common. Similarly, scurvy was increasing at a rapid rate. There were only five cases in 1630, but since 1656 there were up to seventy mortal cases per year.[115] Not only were some diseases increasing in virulence, others were changing their nature. In a long section, Nedham discussed agues, fevers, the French pox, hysteria, convulsions, smallpox, and measles. All of his data tended to support the view of Van Helmont that the seeds of disease were growing stronger.[116]

---

[109] Marchamont Nedham, *Medela Medicinae, A Plea for the Free Profession, or a Renovation of the Art of Physick, Out of the Noblest and most Authentick Writers* . . . (London: Richard Lownds, 1665), p. 7.

[110] *Ibid.*, p. 15.

[111] *Ibid.*, p. 16.

[112] *Ibid.*, p. 255.

[113] *Ibid.*, p. 249.

[114] *Ibid.*, pp. 42–44.

[115] *Ibid.*

[116] *Ibid.*, p. 158.

The fact that the most dangerous diseases were becoming more rebellious[117] indicated that they had been altered from a milder form in antiquity. The only reasonable conclusion was that there was a need for "new Doctrines, new Methods, and Rules of Creation agreeable to the new frame of human nature, and to the new Phaenomena of Diseases."[118] If we succumbed to the demands of the Galenists "to take away *Mercury*, *Antimony*, and *Vitriol*, . . . you leave the Armory of Physick reproachfully weak."[119]

What was needed for medicine was a new "Experimental Philosophy." Quoting the medical disputations of Sylvius, Nedham noted that

> I cannot but wonder . . . that some Physicians, who are eminent for learning, should be so absurd, as not to be ashamed even now openly to profess, that all things which may make for the perfection of Physick were known to any one or more of the Antients, although they are discovered to be false by every mans experience.[120]

Had not Descartes led a solitary life for twenty-five years devoted to experiment? The result had been "a new Philosophy to lead men to Physick."[121] And Thomas Willis, in the preface to his work on fevers, told his readers of the need to depart from the ancients.[122]

Nedham then turned to the subjects that must be excluded from medicine. Surely here we must start with the four elements and the four qualities.[123] With them will go the temperaments, the humors, and the complexions. These were to be replaced with the five chemical principles as discussed by Willis in the *De fermentatione*: Spirit, Sulphur, Salt, Water, and Earth.[124] These are demonstrated to our senses through chemistry, which is essential for the medical doctor. To these principles must be added the concept of the seed as discussed by van Helmont.[125] Nedham paused briefly in his discussion to note that

---

[117] *Ibid.*, p. 203.

[118] *Ibid.*, p. 204.

[119] *Ibid.*, p. 78.

[120] *Ibid.*, p. 233.

[121] *Ibid.*, pp. 235–36. See Davis, *op cit.*, pp. 101–7 for a discussion of Descartes' use of fermentation. This surely made his work of interest to chemists.

[122] Nedham, *op. cit.*, p. 237.

[123] *Ibid.*, pp. 243–45, 257.

[124] *Ibid.*, pp. 271–73.

[125] *Ibid.*, p. 279.

> The difference betwixt the Scholastick men and Us, is shortly this; Their Doctrin is founded upon *Elements*, *Qualities*, *Contrarieties*, *Mixtures*, *Humors*, and old Medicins; We insist upon the *Internal Seminal Principles* of Diseases, the *manna of their Production*, their *Causes of Adaptation*, their *Affinities*, *Consents*, *Diversifications*, and proprieties, with a perpetual improvement of *Medicinal Inventions*, by the Preparations, Exaltations, and Appropriations of the Sublimer and more refined sort of remedies . . .[126]

Other differences between the two systems singled out by Nedham included the Galenic system of three seats of digestion versus the six digestions of van Helmont,[127] and the importance placed on the ferment by the chemists.[128] Here he cited the "noble Mr. *Boyle*" to the effect that scarcely any disease could be understood without a knowledge of the ferment.[129] Van Helmont's *archeus* was credited with being the "director general of all works of Generation, yea and Corruption also."[130] It was endowed with the faculty "to put in execution this or that Function for the service of the Body, but when it is infected, or enraged, then to raise Tumults, Discrasies, and Disorders, to the detriment of the Body: which exactly suits with the Notion of *Helmont*."[131]

Again Nedham followed van Helmont in rejecting the prevalence of bloodletting in medical practice[132] and he urged the use of chemically prepared medicines because they are not liable to corruption and much better for patients.[133] He added that if the Galenists would not convert to these remedies then the Apothecaries had reason to leave them.[134] One of the problems was that many people thought that the chemical physicians only prescribed metallic and mineral remedies

> whereas none deal more in *Vegetables* and parts of *Animals* than they do, for the making of Medicins; only *Mineral* Medicaments are used

[126] *Ibid.*, p. 282.
[127] *Ibid.*, pp. 284–85.
[128] *Ibid.*, p. 286.
[129] *Ibid.*, p. 293. See also p. 512.
[130] *Ibid.*, p. 297
[131] *Ibid.*, p. 301.
[132] *Ibid.*, pp. 390–96.
[133] *Ibid.*, p. 502.
[134] *Ibid.*, p. 433.

upon occasion, and may be with as much safety as the others, at the discretion of such Physicians as take care of the preparation.[135]

Some of Nedham's authorities have already been noted. Although he seldom referred to Paracelsus, he cited Robert Fludd at length on sympathetic cures,[136] Descartes because he had founded a new philosophy based on experiment,[137] and Francis Bacon for his *Advancement of Learning*.[138] He felt that Harvey was one of the great figures of his age, and Thomas Willis, largely because of the *De Fermentatione*, was "the Ornament of our Nation next to immortal *Harvey* . . . one that hath made himself a Physician indeed and Philosopher by Fire."[139] Along with some other chemists–although not all–he recommended the *De abditis rerum Causis* of the sixteenth-century French physician, Jean Fernel, who "had so much good Philosophy in his head, as had he lived in this Age to see what we see, would quickly have brought him off to an imbracement of the profession of *Chymical Physick*."[140]

Clearly van Helmont loomed large in Nedham's thought. Van Helmont was "profound" and "most acute," "the reading of whom I can never sufficiently *commend to our Countrymen*, now that he is printed in English; but know, 'tis not a slight reading will serve the turn: Ten times over is too little, though a man should have the help of a Tutor to direct him."[141] Van Helmont's disciple, Oswald Grembs, is quoted a number of times, as is Johannes Zwelfer, whose annotations on the *Pharmacopoeia Augustana* (1653 and later) reflected Helmontian concerns.[142] We will meet Zwelfer again as he delivers his attack on Otto Tachenius. Sylvius is cited with approval and the French chemist, Nicholas Lefèvre, who set up a chemical laboratory in the Palace at

[135]*Ibid.*, p. 502. It should be noted that the chemical texts of the seventeenth century were always divided into mineral, vegetable, and animal preparations–not metallic and mineral substances alone.

[136]*Ibid.*, p. 122.

[137]*Ibid.*, pp. 235–36.

[138]*Ibid.*, p. 7.

[139]*Ibid.*, p. 237.

[140]*Ibid.*, p. 440. On Jean Fernel (1497–1558), see the article by Ragnar Granit in the *Dictionary of Scientific Biography*, *4* (New York: Scribner's, 1971), pp. 584–86, as well as C. S. Sherrington, *Endeavour of Jean Fernel* (Cambridge: Cambridge University Press, 1946). In the *De abditis rerum causis* (1548), Fernel wrote of sympathetic action of medicines, the philosopher's stone, and transmutation. See Debus, *French Paracelsians*, pp. 13–14.

[141]Nedham, *op. cit.*, pp. 156, 239.

[142]*Ibid.*, p. 16.

Whitehall for Charles II, is quoted frequently.[143] Robert Boyle has many pages in the *Medela medicinae* taken from the *Usefulness of Experimental Philosophy*. However, Nedham did not mention the attack on the principles found in *The Sceptical Chymist*, which he preferred to ignore.

NEDHAM AND HIS OPPONENTS: SPRACKLING, TWYSDEN, AND CASTLE

Nedham's book was answered in short order by Robert Sprackling, (d.c. 1670), a "Candidate Fellow" of the College of Physicians. His *Medela Ignorantiae*, licensed 24 February 1664/1665, supported the ancient medical authorities as well as the traditional four elements, humors, and temperaments. He began with a defense of the *Aphorisms* of Hippocrates, taking note that although Nedham (whom he knew only by his initials) had attacked them, even "*Paracelsus* taketh notice of the verity and usefulness of several of those *Aphorisms*."[144]

The question of chemical medicines loomed large in Sprackling's account. M.N. was to be censured for suggesting that the apothecaries turn entirely to chemical medicines. But it should not be thought that the fellows of the College favored Galenicals alone. None of them rejected chemical preparations; they only sought to restrain those empirics and charlatans who were willing to sell all kinds of dangerous substances.[145]

> Let the gallant Art of Chymistry enjoy its due honour, for the delightful variety of operations, pleasantness and efficacy of Remedies, clear discovery of the nature of Vegetals, Animals and Minerals, correction of deleterial, and otherwise unsafe Medicines; facility, and in many cases, cheapness of cure. But . . . the *Galenical* . . . must not be cast out of doors, so long as they do good service . . .[146]

---

[143] *Ibid.*, p. 250.

[144] Robert Sprackling, Doctor of Physick, Coll. Lond., *Medela Ignorantiae or, A Just and plain Vindication of Hippocrates and Galen from The groundless Imputations of M. N. wherein The whole Substance of his Illiterate Plea, intituled Medela Medicinae is occasionally considered* (London: W. G. for Robert Crofts, 1665), pp. 21–22.

[145] *Ibid.*, p. 30.

[146] *Ibid.*, pp. 31–32.

In short, the traditional Galenic medicines remained of value even if chemicals were at times to be preferred.[147] Sprackling urged caution in the reading of authors. In particular he was certain that Boyle's *Usefulness of Experimental Philosophy* has been misinterpreted in the *Medela medicinae.*[148]

Sprackling was surely not wholly opposed to the rival philosophies of his day. If on the one hand he defended the ancients, he realized on the other that the chemical doctrine of principles had value in the division of "composites into their constituent and separable parts." But this philosophy "may be still completed by the atomistical improvers . . ."[149] Nevertheless, until it is, we may still profit from the traditional elements and humors.

Sprackling clearly differentiated between Paracelsus and his followers on the one hand, and Tycho Brahe, Galileo, and Descartes on the other.[150] He also recognized that van Helmont attributed "his new Doctrine to private inspiration" rather than observation and experiment.[151] And finally he noted that M.N. had resurrected the attack on Duchesne and Turquet de Mayerne in Paris early in the century as a reason to reject the medical establishment. But surely, Sprackling countered, all medical societies should not be banished because of the "rigour" of the Medical Faculty at Paris in the past. The attitude of the London College of Physicians is seen best in the fact that while Mayerne was condemned in Paris, he was rightly honored in London.[152]

By the spring of 1665 the Plague was in London and the chemical physicians were in the process of organizing their society. However, it is not our purpose to cover all of the literature related to these events, but rather to hold to the tracts directly referring to Nedham's *Medela medicinae.* The next to appear was Edward Bolnest's *Medicinae Instaurata,* which included a preface by Nedham dated 10 May 1665.[153] Bolnest was a friend of Nedham and a fellow organizer of the Society of Chemical Physicians. In his preface Nedham praised his friend and commented that although others may call us "*a Company of* Illiterate Professors, . . . *there is a way to true Physical Learning, which lies*

---

[147]*Ibid.,* p. 33.

[148]*Ibid.*

[149]*Ibid.,* p. 45.

[150]*Ibid.,* p. 156.

[151]*Ibid.,* p. 157.

[152]*Ibid.,* pp. 158–59.

[153]Edward Bolnest, *Medicina Instaurata, or A Brief Account of the true Grounds and Principles of the Art of Physick . . . Also an Epistolary Discourse upon the whole by the Author of Medela Medicinae* (London: John Starkey, 1665), sig. a5^{v}.

quite out of the Common Scholastick Road; and that we must pass through the fire to it, while the fine Fellows are afraid to follow for fear of singeing their Scarlet."[154] At this point Nedham expressed concern about those at the College who say that they are chemists themselves and that there is therefore "*no need of erecting a new Society upon the account of* Chymistry."[155] Indeed, Nedham argued, both Bolnest and he were University-trained and could be members of the Royal College of Physicians if they so desired, but that was not their wish since the only way "to redeem our Profession" was to set "the whole frame of Physick upon a new foot of Operative and Experimental Philosophy," in a word, upon a chemical foundation.[156]

Nedham and other chemists had reason to be concerned about the Fellows' assertion that they supported chemistry. By the early 1660s the College had erected a chemical laboratory with William Johnson in charge. His *Lexicon Chymicum*, issued in two parts in 1652 and 1653, was to become one of the most authoritative chemical dictionaries of the period and it was dedicated to Sir Francis Prujean (1593–1666), who served as President of the College from 1650 to 1654.

In the College laboratory Johnson prepared chemical medicines for the use of the Fellows in practice.[157] With this establishment connection, Johnson entered the raging polemical debate, although he did not aim his *Agyrto-mastik* (1665) specifically at Marchamont Nedham. While he believed that Galen and van Helmont could be reconciled, he felt that there had been a dangerous growth in the ranks of the pseudo-chemists.[158] Rather, he praised the Fellows of the College who had encouraged him.[159] Worse still from the viewpoint of the Helmontian chemists was the fact that Johnson argued that the present high state of this "Noble Art," "is ow'd to them [the Fellows of the Royal College] and if it ascent yet higher, the same hands must lend their assistance."[160] Johnson further spoke of the need of anatomical

---

[154] *Ibid.*, sig. A5^{v}.

[155] *Ibid.*

[156] *Ibid.*, sig. A6^{r}–A7^{v}.

[157] Sir George Clark, *A History of the Royal College of Physicians of London* (3 vols., Oxford: Clarendon Press, 1964, 1966, 1972), *1*, p. 286.

[158] William Johnson, Chymist to the Kings College of Physicians in London, *Αγυρτο-Μαστιε, or some Brief Animadversions Upon two late Treatises; One of Master George Thomsons, Entituled, Galeno-Pale; The other of Master Thomas O'Dowdes, Called The Poor Mans Physician: With a short Appendix, relating to the Company of Apothecaries* (London: T. Mabb for Henry Browne, 1665), sig. A3^{v}.

[159] *Ibid.*, pp. 7–8.

[160] *Ibid.*, p. 8.

studies,[161] and–quite understandably–he insisted that the apothecaries only use medicines obtained from the College laboratory.[162] It is little wonder that Nedham and other Helmontians looked upon this activity of the College with alarm.

To return to the preface to the *Medicina Instaurata*, Nedham went on to attack the author of the *Medela Ignorantiae*, who had called him illiterate.[163] Sprackling was surely one of those who erroneously thought that medicine could maintain its reliance on ancient authorities while adding a few chemical remedies to this time-worn base. In fact, the chemical remedies in the *Pharmacopoeia Londinensis* of the College were quite insufficient and the traditional "method" of the physicians would have to be discarded. The new Chemical Society would be able to furnish apothecaries with all the approved chemical preparations after their new building was completed.[164] This point was important and Nedham returned again to his conviction that a number of deadly diseases were changing radically. This was particularly true of the venereal diseases, which had become considerably more virulent over the past two or three decades. The old approved cures were useless and the Fellows of the Royal College were well aware of this.[165]

Bolnest's work was largely an advertisement for the chemical remedies he employed in his practice–in particular his "Solar Pill." Bolnest referred to a spectrum of chemists from Paracelsus to his own day and, in "An Expostulary Conclusion to my Ingenuous Contreymen, in behalf of Chymistry and Chemical Medicines," he asked why the Galenists complained about their chemical preparations when they "make use of them? Why do they steal our *Extractum Rudii (alias Extract Catholicum Paracelsi)?* Why our *Laudanum? Pilula Alaephany?*"[166] and many others. Indeed, they may pretend to hate our chemicals when they truly know their value.

The Bolnest volume was in turn answered by John Twysden in an *Answer* to the *Medicina Instaurata* (1666), and then, taking on Nedham, he wrote the *Medicina Veterum Vindicata* (1666). Twysden (1607–1688) was educated at University College, Oxford, and took his M.D. at Anger (1646).[167] He had a

[161] *Ibid.*, p. 26.

[162] *Ibid.*, pp. 133–34.

[163] *Ibid.*, Bolnest, *op. cit.*, sig. a2^r.

[164] *Ibid.*, sig. a4^v–a5^r.

[165] *Ibid.*, sig. a5^v.

[166] *Ibid.*, pp. 138–39.

[167] On Twysden, see the article by Norman Moore in the *DNB*, *19*, p. 1331. See also King, *op. cit.*, pp. 154–60.

persistent interest in mathematics and published more in that field than he did in medicine. Nevertheless, he was elected a Fellow of the Royal College of Physicians in 1664 just as the chemical physicians were about to challenge that society.

In his *Medicina Veterum Vindicata*, he complained that the chemists had tried to get the King to sponsor their society.[168] In fact, Twysden argued, many members of the College were skilled in chemistry.[169] In any case, the actual practice of chemistry differs from the writings of Paracelsus and van Helmont, which are "so obscure, that I shall forbear any censure upon them, till this or some other charitable person shall make me clearly understand the meaning of them."[170]

Twysden took exception to the chemists' rejection of the study of anatomy which, he concluded, was as necessary for medicine as mathematics was for astronomy.[171] He rejected atomism,[172] and wrote that both Harvey and Boyle were his personal friends.[173] Like Sprackling, Twysden charged that Nedham had misrepresented the work of Boyle, who surely would not want to be party to the machinations of the chemists.[174]

A less conventional reply to Nedham appeared from the pen of George Castle (ca. 1635–1673), an M.D. from All Souls' College at Oxford (1665) who was later to be elected a Fellow of the Royal Society.[175] His only book, *The Chemical Galenist* (1667) is specifically directed at the *Medela medicinae.* Castle fully appreciated the new discoveries in the sciences and medicine and he wrote that the

> Physiology and Pathology of Medicine should be modelled according to the new discoveries of Anatomy, and the Democriteal and Chymical Principles, yet that many of the Rules, Methods, and Medicines, which more immediately respect the useful and practical part,

---

[168]John Twysden, M.D., Fellow of the Kings College of Physicians in London, *Medicina Veterum vindicata: Or an Answer To a Book, entituled Medela Medicinae*... (London: J. G. for John Crook, 1666), sig. A3^r.

[169]*Ibid.*, sig. a3^v.

[170]*Ibid.*, p. 15.

[171]*Ibid.*, pp. 30–31.

[172]*Ibid.*, p. 93.

[173]*Ibid.*, pp. 47, 114–15.

[174]*Ibid.*

[175]On Castle, see the article by Gordon Goodwin, *DNB*, *3*, p. 1183.

> are still to be retained, and that they are rather more reconcileable to the Modern than they were to the Ancient Hypotheses.[176]

Yet, if this be so, one must avoid the extremes of the militant chemists "who would make the world believe that they [the university-trained physicians] are a company of narrow streight-lac'd men, who onely confine themselves to Books, and never study Nature, and that they stifle and suppress all occasions of improving their Art."[177] In fact, Castle believed that the recent improvements in medicine were due entirely to properly trained physicians, men such as Harvey, Glisson, and Willis.[178]

Castle disagreed with the growing concept of a chemical physiology. Rather, he argued that

> . . . a man is as Mechanically made as a Watch, or any other Automaton; and that his motions (the regularity of which we call Health) are perform'd by Springs, Wheels, and Engines, not much differing (except as to the curiousness of their Work) from those pieces of Clock-work, which are to be seen at every Puppet-play.
>
> He will certainly not rest better satisfied in the Mechanical account of the operations, and diseases of an Animal, than in the *Ens Pagoicum, Cagastricum, Illiastrum, Archeus Relloleum, Chironeum, Evestrum, Yleck, Trarames, Turban, Lessas, Srannan, Perenda, Zenda*, and a thousand such conjuring unintelligible words of the *Chymists*.[179]

It is anatomy that permits us to view man in terms of clockwork. We now view the valves of the heart and the waterworks of the kidneys.[180] This being so, what use are cures compounded of antimony or mercury? To be fair, there is some value in the use of chemistry for pharmaceutical preparations, but chemistry was of little use to medicine "till it fell into the hands of Rational Learned Men; who by adapting it to the Atomical Philosophy have

[176]George Castle, Dr. of Physick, lately Fellow of All-Souls Colledge in Oxon., *The Chemical Galenist: A Treatise Wherein The Practice of the Ancients is reconcil'd to the new Discoveries in the Theory of Physick* . . . (London: Sarah Griffin for Henry Twyford, 1667), sig. A4^r.

[177]*Ibid.*, sig. A6^r.

[178]*Ibid.*, p. 3.

[179]*Ibid.*, pp. 5–6.

[180]*Ibid.*, p. 8.

made excellent use of the Analysing of Bodies."[181] We need not be surprised to see that when Castle referred to Willis, he cited his anatomical description of the brain rather than his chemical speculations in the *De fermentatione.*[182] And when he referred to Robert Boyle, it was to *The Sceptical Chymist* against the unnecessary obscurity of some chemists rather than *The Usefulness of Experimental Philosophy.*[183]

Turning to Nedham's reference to the changes to be found in diseases in recent decades, Castle ascribed this largely to the fact that we are now more aware of some diseases and, because of this, they attract more attention when they occur.[184] And in the case of much longer periods of time, we are faced with the fact that man's diet has changed and that because of this we are softer and more prone to react violently to the same diseases of the past.[185]

In the end Castle returned to his opening thoughts. We should incorporate the latest discoveries into medicine, but

> care must be taken, that we imitate wise and thrifty Builders, who, in raising a new house in the place of an old one which they have pull'd down, may make use of many of the old substantial Materials, some of which are often much the better for their age.[186]

Other references to the *Medela medicinae* may be found in succeeding years, such as Henry Stubbe's work on the *Sweating Sickness* in 1671,[187] but by the late 1660s the chemical physicians no longer posed the threat to the College of Physicians that they had in 1665. Marchamont Nedham did return to a defense of the chemists in 1675, however, when he prepared a long preface to an English translation of the first book of Sylvius's *Praxeos Medicae Idea Nova.* This translation had been prepared by Richard Gower, an English stu-

[181] *Ibid.*, p. 10.

[182] *Ibid.*, p. 44.

[183] *Ibid.*, p. 16.

[184] *Ibid.*, p. 44.

[185] *Ibid.*, p. 32.

[186] *Ibid.*, p. 196.

[187] Henry Stubbe, *The Lord Bacons Relation to the Sweating-Sickness Examined, in a Reply to George Thomson, Pretender to Physick and Chymistry. Together with a Defence of Phlebotomy in General, and also particularly In the Plague, Small-pox, Scurvy and Pleurisy. In opposition to the same Author, and the Author of Medela Medicinae, Doctor Whitaker, and Doctor Sydenham . . .* (London: Phil Brigs, 1671).

dent who had matriculated at Leiden in 1670 and had studied under Sylvius.[188] It is of interest to note that Gower dedicated his work to Robert Boyle. Nedham's preface to this work is of special interest because Sylvius was not one of his primary authorities when he wrote the *Medela medicinae.*

In his preface, Nedham sought to establish the point that Sylvius was, above all, a dedicated chemist. For Nedham and other Helmontian chemists of his day anatomy posed a problem. As we know, Sylvius was a skilled anatomist as well as a chemist, but like other Helmontian chemists, Nedham attacked the study of anatomy. He complained that too many anatomists simply boasted of the parts of the body that they had discovered, knowledge that "is of little use in managing the Cures of most Internal Diseases."[189] Far too many students waste their time on anatomy of the human body when they should be chemically anatomizing natural things that will lead to improvements in the *materia medica.*

> Alas! what a world of ground lies waste in the Field of Nature! how comly would it be, to see them spend their time in making further inquiries into Herbs and Roots which contain innumerable other virtues than yet are known. Had we not been more beholden to Women and Rusticks, than to the Men of Literature, very little more had been known of them now, than was a thousand years ago.[190]

Students then should spend less time on anatomy than on acids, salts, and other chemical properties "wherein our Sylvius here hath pointed out part of the way, those things being by experience found to be the real principal distrubers of Humane Bodies."[191]

But Sylvius had established his points through anatomical evidence. How could this be supported? For Nedham the answer was that his medical

---

[188]According to R. W. Innes-Smith, Gower was from Spittle, Newcastle-on-Tyne and matriculated at Edinburgh University in 1665. He translated a work by Sylvius on children's diseases in 1682. *English-Speaking Students at the University of Leyden*, with a Foreword by John W. Comrie (Edinburgh: Oliver and Boyd, 1932).

[189]Franciscus de la Böe, Sylvius, *A New Idea of the Practice of Physic; . . . The First Book . . . All cleard by Anatomical Experiments; As also by their Cures. Whereto is prefixed a Preface written by Dr. Mar. Nedham, Translated faithfully by Richard Gower, formerly Student under the Author* (London: Brabazon Aylmer, 1675), from the prefatory "To the Reader by Nedham," sig. b8^{v}.

[190]*Ibid.*, sig. b9^{v}.

[191]*Ibid.*, sig. c1^{v}.

lectures were more for traditional "Pomp and Ceremony" than for real use. Thus, the anatomical work of Sylvius could be dismissed. He was important

> because he raised most of his Doctrines from Chymick Experiments, the great things that enlighten us in the Art of Curing, and verified them with pleasant and speedy Cures, by joining the use of Plants and Minerals together, both Chymically prepared, which hath rendred him worthy the reading and imitating.[192]

In short, Sylvius was important for his chemical contributions. The fact that he frequently referred to and discussed human anatomy was to be dismissed as a necessary requirement associated with his Leiden professorship.

CONCLUSION

In the writings of Sylvius, Willis, and Boyle from the 1650s and 1660s, we see a powerful influence of current chemical thought in medicine. This is surely due in part to the publication of the collected works of van Helmont in 1648, which intensified the continuing interest in Paracelsian thought. With Sylvius we see a greatly increased emphasis on chemical explanations of physiological processes. This is particularly evident in his interest in acid-alkali reactions as well as his use of fermentation as a basis of explanation. At the same time, we see the study of human anatomy joined with chemistry in the works of Sylvius–a union that was not approved of by all chemists.

This union of anatomy and chemistry is even more pronounced in the work of Thomas Willis, who was surely one of the most gifted anatomists of the seventeenth century. But for chemists who only approved of "anatomy by fire" there was his *De fermentatione* to cite. In this early work (1659), there is little reference to human anatomy, but a rich treasury of chemical thought. Here Willis turns to ferments and fermentation as a basis for understanding both man and the great world. And here also he supports the chemical principles as a foundation for natural philosophy. Although Willis did not place the same emphasis on acids and alkalis that we find in Sylvius, he did support the use of chemically prepared remedies in his vast *Pharmaceutice Rationalis.*

As for Robert Boyle, when he was a young man he was drawn to the work of van Helmont and in many ways his work from the 1650s is deriva-

---

[192] *Ibid.*, sig. c2^{v}–c3^{r}.

tive of the Belgian's. He was critical of the traditional humors of Galenic medicine and he reflected van Helmont's views on the macrocosm-microcosm analogy. He certainly believed in the use of chemical remedies and he wrote of the need for a chemical understanding of fermentation if we are to understand disease. And it has been shown conclusively that his critique of the elemental systems of the Aristotelians and the Paracelsians in *The Sceptical Chymist* was openly drawn from van Helmont.

While these three figures, Sylvius, Willis, and Boyle, do not begin to cover the spectrum of interest in chemical medicine in the mid-seventeenth century, they do help to explain the debate that followed the publication of Marchamont Nedham's *Medela medicinae* in 1665. Nedham, as an Helmontian chemical physician, was active in the move to establish a society of chemical physicians in that year. In common with his fellows, he demanded a complete reform of medicine on the basis of chemistry. He rejected the study of human anatomy and turned rather to the chemical laboratory that would present us with new remedies. As might be expected, he supported the chemical principles as a basis for his medicine and natural philosophy.

Nedham considered Sylvius, Willis, and Boyle to be major authorities, but they were to be used selectively. Boyle's *Usefulness of Experimental Philosophy* was quoted at great length, but not *The Sceptical Chymist*, even though both works were Helmontian. His lack of interest in the latter surely resulted from Boyle's attack on the chemical principles, a fault not to be found in the *Usefulness of Experimental Philosophy*. In regard to Willis, Nedham praised the *De fermentatione* at length, but chose to ignore his contribution to anatomy. He had the same problem with Sylvius, who was quoted in regard to his views on the necessity of chemistry for a proper understanding of natural science. However, there was again the problem of anatomy. Here Nedham met the problem directly and stated simply that Sylvius had to lecture on anatomy because he was a professor of medicine and that was required of him.

The replies to Nedham by Sprackling, Twysden, and Castle refer to Sylvius, Willis, and Boyle from different points of view. In general they agreed that Nedham did not fully understand Boyle and they all supported the need for anatomical studies in medicine. Perhaps most galling to Nedham and his fellow chemists was the fact that these authors argued that they had accepted chemistry and that they were the ones responsible for its continued advancement in medicine. They had established a chemical laboratory at the Royal College of Physicians and they had even appointed a distinguished chemist, William Johnson, to be its head. In 1665 Johnson had even entered the polemical war to support the Fellows in their battle with the Helmontian chemists.

In the end, the most interesting opponent of Nedham may have been George Castle, who wrote of the value of the new scientific discoveries and even of the use of chemistry for the discovery of new medicines. But Castle also added that man's body is best seen as clockwork. His work is an early indication of the forthcoming debate between the chemists and the mechanists in medicine. But although this was on the horizon, we must first continue with the development of the acid-alkali system in medicine.

CHAPTER 4

Acids and Alkalis: The Chemistry of Hippocrates

> . . . *Natural Philosophy (now styled* Chymistry*) according to Hippocrates, is a most Ancient Science, being derived from the Greek words,* 'Als, *signifying* Salt, *and* Cheo *to Melt, as much as to say,* The Art of melting Salts, *And 'tis rightly called so.*
>
> Otto Tachenius, *Hippocrates Chymicus*, trans. I.W. (1677)[1]

There is little doubt that chemistry seemed to pose a major problem for many physicians in the late seventeenth century. More than a century of debate had produced a series of court battles and numerous polemical tracts centered on this controversy. The debate between the chemical physicians and the fellows of the Royal College of Physicians in London in 1665 represents only one example of a continuing confrontation that affected the medical profession throughout Europe.

At this time, however, the debate was still far from over. The influence of van Helmont, followed by Sylvius and Willis, had placed a new emphasis on the use of chemistry as a key to understanding life processes through the concepts of fermentation and the acid-alkali dichotomy. Thus, along with the continued pressure for the acceptance of chemically prepared medicines, we see an increased interest in a chemical physiology. This is evident in the work of Otto Tachenius, in which there is a new effort to unite a medicine

---

[1]Otto Tachenius, *Hippocrates Chymicus, Discovering the Ancient foundations of the late Viperine Salt with his Clavis there unto annexed*, trans. I. W. (London: Nath. Crouch, 1677), sig. B3^{v}.

founded on the duality of acids and alkalis with Hippocratic tradition. And if these were major areas of concern in the late seventeenth century, traditional areas of conflict remained evident as well. This is readily apparent in the literature of the period.

CHEMISTRY IN MEDICAL THEORY

As we look at the scientific and medical literature of the third quarter of the seventeenth century, we constantly find the Chemical Philosophy presented as a major approach to nature. This has been seen already in debates centered on Paracelsian and Helmontian natural philosophy. And it is evident again in the work of Jean-Baptiste Du Hamel (1623–1706), who is best known as the first permanent secretary of the Académie Royale des Sciences (1666–1697).[2] A student first at Caen and then Paris, Du Hamel later taught philosophy at Angers. He was known both for his work in anatomy and astronomy, but because of his broad interests it would be difficult to associate him with any one science. This is particularly true of works he wrote in the early 1660s, in which he discussed the current state of the sciences. Here our interests center on his *De consensu Veteris et Novae Philosophiae* published in 1663.[3]

For Du Hamel there was truth in all the major philosophies, but this was not readily evident because the followers of each sect were determined not to borrow from others, convinced that theirs excelled in everything. To be sure, Aristotle surpassed the others in logic, but the chemists had gone farther in their explanation of the principles of natural bodies, while Descartes was the outstanding author on motion. However, the chemists rejected the general notions and terms of the Aristotelians, preferring to invent new ones and to confuse their readers with an entirely new jargon. And the Peripatetics would accept nothing unless it was to be found in Aristotle. The Cartesians were no better than the others because they rejected all other philosophies and had developed an entirely new physics and metaphysics.[4]

Thus we find whole systems of thought being developed on the basis of the opinions of a particular sect. Du Hamel continued that the Platonists

---

[2]See the biographical sketch by Pierre Costabel in the *Dictionary of Scientific Biography*, vol. 4 (New York: Scribner's, 1971), pp. 221–22.

[3]See Allen G. Debus, *The French Paracelsians* (Cambridge et al.: Cambridge University Press, 1991), p. 149.

[4]Jean-Baptiste Du Hamel, *De Consensu Veteris et Novae Philosophiae Libri Duo* (Paris: Apud Carolum Savreux, 1663), sigs. eiii^{r}–ev^{r} ("Ratio operis").

excelled in natural theology; they pretended that all philosophy consisted in the contemplation of God and that there was no need to consider secondary causes. The disciples of Democritus and Epicurus explained the principles of natural bodies by the doctrine of atoms and they paid little attention to anything else. As for the Aristotelians, they accommodated their natural science to their metaphysical speculations, while the Cartesians had reduced all to geometry and mathematics. Du Hamel argued that in reality the ideal philosophy should borrow the best from all sects.[5]

The volume was divided into two main sections; the first described the views of the Platonists, the Aristotelians, the Epicureans, and the Cartesians regarding the internal and external principles of bodies, whereas the second discussed the views of the chemists. "He examines all their opinions; he compares one with the other; he shows how they differ and how they might be conciliated. After this he chooses the one which he judges the most like to be true."[6] The reviewer in the *Journal des Sçavans* (referring to the second edition of 1669) was impressed by the understanding of Du Hamel. "And particularly one finds here a very exact abridgment of the physics of Mons. Descartes, while all of chemistry is also explained clearly and in few words."[7]

There is little doubt that Du Hamel was well versed in the chemical literature of the period. In the *De consensu*, he discussed the transmutation of the four elements into one another as well as the Paracelsian principles. In this exposition of the elements, he gave examples by way of the distillation of wine, wood, and vitriol.[8] In a later book, the *De corporum affectionibus cum manifestis tum occultis* (1670), he discussed recent observations, such as van Helmont's willow tree experiment and the use of the thermometer for temperature measurement.[9] Throughout all of this one finds the chemical philosophy presented as a subject of major concern for natural philosophers.

Another approach to the chemical philosophy may be seen in the work of Daniel Duncan (1649–1735), who was born at Montaubon in Languedoc, where his father, Peter, was Professor of Physic.[10] He studied at Montpellier, where he took his medical doctorate in 1673. He was appointed physician-

[5]*Ibid.*

[6]Review of the *De Consensu* in *Le Journal des Sçavans* (1669), pp. 44–46 (45).

[7]*Ibid.*, p. 46.

[8]Du Hamel, *De Consensu* (1663), Liber Secundum, pp. 183–85.

[9]See the review of Du Hamel's *De corporum affectionibus cum manifestis, tum occultis, libri duo* (Paris: apud M. Le Petit, & S. Michallet, 1670) in *Le Journal des Sçavans* (1670), pp. 14–15.

[10]See Debus, *French Paracelsians*, pp. 134–36.

general to the army at St. Omer (1677). A Protestant, he left France after the Revocation of the Edict of Nantes (1685). In the following years he lived in Switzerland and Germany–first at Cassel (1699) and then Berlin (1702), where he was appointed physician to Frederick I of Prussia. Eventually he settled in England (1714), where he lived for the remainder of a long life.

It is likely that because of his training at Montpellier Duncan became an advocate of iatrochemistry. For him the mechanical and the chemical philosophies together were a single new philosophy ("la Philosophie nouvelle").[11] Because of its light, the enigmas, the fables, and the barbarous terms of the ancients had been driven away so that now some of the darkest corners of nature might be viewed. "One could say that mechanics and chemistry are like two stars that show the truths of nature in this great day."[12] Indeed, his book was titled *La Chymie Naturelle ou l'Explication Chymie et Mechanique de la Nourriture de l'Animal* (1681).

A dedicatory sonnet by Mr. D. Bilieres compared Duncan with Descartes, to the detriment of the latter:

> Selon Descartes, qui rafine
> Sur tous les Esprits les plus fortes
> Dans la conoissance des corps,
> Nôtre corps n'est qu'une machine.
>
> C'est ainsi qi'il le determine,
> Contemplant ses divers ressorts
> Disposez dedans & dehors
> D'une façon toute divine
>
> DUNCAN, tu passes plus avant
> Que cét ingenieux Sçavant,
> Dont ou revere la memoire;
>
> Quãd tu dis que Dieu de sa main
> A bâte dans le corps humain
> Vn merveilleux Laboratoire[13]

[11]Daniel Duncan, *La Chymie Naturelle ou L'Explication Chymique et Mechanique de la Nourriture d l'Animal* (Paris: Laurent d'Houry, 1683), sig. *v^{r}.

[12]*Ibid.*, sig. *v^{v}.

[13]*Ibid.*

Duncan was clearly interested in understanding the human body and there is little doubt that his interests centered more on chemistry than on mechanics. He might echo Descartes in referring to the animal body as a hydraulic machine and in comparing bodily circulations with tourbillons,[14] but for the most part Duncan understood physiological processes as chemical ones. Surely "if Plato had reason to say that God made nothing without following the rules of geometry, one might add that he produced nothing without those of chemistry."[15]

Duncan wrote of fermentations, filtrations, and circulations in the body and repeatedly offered examples from chemistry to explain physiological observations. Believing in the chemical principles, he argued that superfluous sulphur was separated chemically from the blood in the liver and this, when mixed with the chyle, caused the reddening of the blood. Similarly, sulphur was responsible for the yellowing in jaundice.[16] That these and other reactions occur is due to the heat of the body. "Fire is the great instrument of natural as well as artificial chemistry. And as each artificial operation requires a certain degree of fire, so too the preparation of each humor which nourishes the body requires a specific heat."[17] In short, "the animal body is like one of those marvelous furnaces in which one can make at the same time a great number of different operations. Each organ is a vessel in which a natural chemistry prepares its proper matter."[18] And indeed, "if chemistry is so necessary for our knowledge of the small world, it is no less so for our knowledge of the great world. It would be no more difficult to show that chemistry is required for the production of minerals, metals, animals and all the phenomena that appear in the universe."[19]

It is clear that Duncan pictured the macrocosm also as a great chemical laboratory.

> All the world is a vast laboratory, of which God is the artist, the sun is the great furnace and fire all together: the spirits, the salts, the sulphurs, the phlegm and the caput mortuum of which all the earth is filled, are the subject or the matter on which this fire acts. The trees

---

[14] *Ibid.*, p. 27.

[15] *Ibid.*, sig. **ii.

[16] *Ibid.*, sig. *viii$^{r\&v}$.

[17] *Ibid.*, sig. **i^{r}.

[18] *Ibid.*, sig. **i^{v}.

[19] *Ibid.*

> and the other vegetation are as the vessels in which these principles sublime, ferment and rectify themselves through diverse circulations, filtrations and cohobations, in order to form in the end a very pure juice which is the matter of fruits.[20]

Duncan thought of his work as an answer to those who considered chemistry to be an art contrary to nature. He felt that they looked on chemical laboratories with disdain, but that they would learn in time that their own bodies were laboratories.[21]

La Chymie Naturelle was a theoretical text of questionable value for the practicing physician. And yet, we need hardly be surprised to find that Duncan felt that a knowledge of chemistry was required in medical practice. He pointedly noted that those who complained of the violence of chemical remedies were those who defended the use of the surgeon's knife.[22]

In the second part of his work, Duncan compared the explorers of the earth to the physician who explored the microcosm:

> . . . anatomy is the art of this voyager. In his voyages he regards the solid parts as land, the humors which moisten them as the sea of the microcosm, the large vessels as the rivers and streams and the small ones as brooks. There are even certain currents which merit the name of torrents or floods because they flow only during a short space of time.[23]

The last image explained female menstruation, in which the vagina acted as a grand canal in the macrocosm.[24]

The anatomy of a woman could be compared with the great world, but it did not fully explain the reason for menstruation. Here he turned to the Bible. There is little doubt that when it was created the blood of Eve was as pure as that of Adam.[25] But Eve sinned by taking the apple and a disastrous ferment resulted that corrupted her blood. This was divine justice and a fitting punish-

---

[20] *Ibid.*, p. 23.

[21] *Ibid.*, sig. **ii^{v}.

[22] *Ibid.*, sig. xxiii$^{r\&v}$.

[23] Daniel Duncan, *Seconde et Troisieme Partie de la Chymie Naturelle, ou L'Explication Chymique et Mechanique De l'Evacuation particuliere aux Femmes, & de la Generation* (Paris, printed at Montaubon: Laurent d'Houry & Daniel Horthemels, 1687), sig. Aiiii^{r}.

[24] *Ibid.*, sig. Aiiii^{v}.

[25] *Ibid.*, sig. Bii^{r}–Biii^{v}.

ment for her crime. Her blood, previously pure, was now disturbed with a violent fermentation and disorder resulting in a monthly evacuation.

> Thus one can prove by divine authority that this evacuation followed the rebellion of the first woman since she did not give birth in a state of innocence, seen through the lack of menstrual blood which prepares the womb for generation. "You will give birth with pain" is a judgment that the Judge of the world pronounced without doubt against a criminal Eve, and the Justice of God does not permit Him to condemn to this punishment an innocent person.[26]

CHEMISTRY IN MEDICAL PRACTICE

Du Hamel and Duncan were clearly interested in theoretical matters, but, as always, most physicians were concerned primarily with medical practice. Thus, James Thompson in an Helmontian apology (1657) discussed cures for fevers, the stone, and the plague–all subjects of special interest for Helmontian physicians. As one might expect, Thompson was opposed to blood-letting because

> (according to the Scripture likewise) the life glideth in the bloud; however, this be taken away largely together, it cannot be done without prejudice to the life also. For the perpetual intention in Nature, in the cure of fevers, is to do it by sweatings; and therefore also the fits of fevers, do for the most part end by sweatings.[27]

Like van Helmont, Thompson cautioned against a mathematical approach–here going from one extreme to the other without a consideration of the mean. Indeed, "the Mathematicks prove blood-letting alwaies hurtfull."[28]

Simon Partliz (Partlicius) von Spitzberg had earlier published his *Medici systematis harmonici* (1625), which was translated into English in 1654 as *A New Method of Physick.* Here he offered to compare the medical practices of

[26] *Ibid.*, sig. Ci^{v}.

[27] J. T. [James Thompson] Esq., *Helmont Disguised: or The Vulgar Errours of Impericall and unskilfull Practisers of Physick confuted. More especially, As they concern the cures of Feavers, Stone, Plague, And other Diseases. In a Dialogue between Philiatros and Pyrosophilus. In which the chief Rarities of Physick is admirably discoursed of* (London: E. Alsop for N. Brook, and W. Leybournse, 1657), p. 104.

[28] *Ibid.*, pp. 104–5.

Paracelsus and Galen. In a brief history of medicine, he traced the subject from Adam to Hermes and Aesculapius and then on to the Hebrew prophets, the Greeks, and the Arabs. "The Alchymists of our times oppose themselves to these, and will be called Paracelsians . . ."[29] Partliz clearly sided with the Paracelsians and he drew heavily on Paracelsian texts to discuss the double body of man, "the one Natural, Elementary, visible and tangible . . .: the other Invisible, Insensible, deduced from the Coelestial Influence of the Stars . . ."[30] He was one of the few authors to describe in Paracelsian terms the two types of disease seeds (Iliastric and Cagastric) and the five beings of disease described in the *Paramirum*.[31]

Above all, the medicine of Partliz was dominated by medical chemistry or alchemy. "The Art of Alchymie is certaine and infallible. For it propounds two scopes to it selfe. Either to transmute Mettals, or, To make profitable and effectual Medicines. Both of them it performeth with great success and praise."[32] Partliz knew the traditional alchemical literature and he paraphrased the Emerald Table of Hermes,[33] but at the same time he clearly viewed alchemy primarily as a medical subject. Here the most practical aspect related to pharmacy.

> Although when you look upon one of *Galens* Apothecaries Shops, you will see fine painted Boxes and curious pots, that it would dazle your eyes to look upon them, they are so finely painted, That if there were a paradice upon Earth you would think it were there: yet in the inside is nothing but filth and the very Carkeises and Dung of all Medicines. Their common Medicines are grosse, crude bodyes, the vices of the simples are in, as well as their vertues, yea so many simples are in one Composition, that they hinder one anothers operation, and therfore how can they ease the sick without calling the help of an Alchymist to resolve, separate, and exhale what is obnoxious, thereby producing the hidden Natures of things for use . . . Also the

---

29 Simeon Partlicus ([Partliz]) Von Spitzberg, *A New Method of Physick: Or; A Short View of Paracelsus and Galen's Practice; In 3 Treatises. I. Opening the Nature of Physick and Alchymy. II. Shewing what things are Requisite to a Physitian and Alchymist. III. Containing a Harmonical Systeme of Physick*, trans. Nicholas Culpeper, Gent. Student in Physick and Astrologie (London: Peter Cole, 1654), p. 33.

30 *Ibid.*, p. 423.

31 *Ibid.*, p. 501.

32 *Ibid.*, p. 51.

33 *Ibid.*, p. 424.

> Alchymist searcheth after the strength and Temperature of things, the Causes and Originall of their actions, and by Mediation of a certain pure body sets the very species and forms before your eyes.[34]

Thompson and Partliz relied on the tenets of chemical medicine for practical results. Christopher Irvine (Irving) (fl. 1638–1685) did so no less with his emphasis on curing by sympathy. Originally a royalist who had been ejected from Edinburgh University with the defeat of Charles I, Irvine turned to medical practice in Edinburgh. After renouncing his allegiance to the Crown, he became a surgeon to General George Monck and after the Restoration continued in that post to the horse guards in London.

Irvine's only published work was his *Medicina Magnetica: or, The rare and wonderful Art of Curing by Sympathy,* published in Edinburgh in 1656 with a dedication to Monck. Here he pictured a cosmos animated with a world soul in which natural things operated through a vital spirit.[35] The seeds of all things contained this spirit and they germinated through fermentation.

> The *universal vital spirit* coming down from *Heaven*, pure, clear, and uncontaminate, is the father of the *particular vital spirit*, which is in every thing: for, it *procreates* and *multiplies* in it the *body*; from whence *bodies* borrow the power of *multiplying* themselves. . . . He that knoweth how to infuse the propitious *Heavens* or *Sun* into *things*, or into the mixture of *things*, may perform wonders, and hereupon depends all *magick operations*.[36]

For Irvine this is a key to medicine since the true physician may apply the universal spirit to any particular substance, resulting in a fermentation that "may miraculously increase things in vertue, and power" to reduce, "appease and settle *Natures tumults*."[37] The desired result of this operation is

---

[34] *Ibid.*, p. 9.

[35] C. de Iryngio [Christopher Irvine], Chirurgo-Medicine in the Army, *Medicina Magnetica: or, The rare and wonderful art of Curing by Sympathy: Laid open in Aphorisms; Proved in Conclusions; And digested into an easy Method drawn from both; Wherein the Connexion of the Causes and Effects of these strange Operations, are more fully discovered than heretofore, All cleared and confirmed, by pithy Reasons, true Experiments, and pleasant Relations. Preserved and published, As a Master-piece in this Skill* [1656 (Epistle dedicated Edinburgh 3 June 1656)], pp. 1–2.

[36] *Ibid.*, pp. 4–5.

[37] *Ibid.*, pp. 7–8 (Aphorism 52).

the prolongation of life.[38] In short, "*The universal Medicine is nothing else, but the spirit multiplied upon a due subject.*"[39]

In his second section, Irvine asserted that the soul of every body works beyond the actual body through the emission of corporal beams that exist in particulate form.[40] Thus a healthy person may beneficially affect a sick one. ". . . [A] wise man will take a special heed to living and conversing with sick people; for a *body* so disposed, doth more greedily draw to it self those *beams*, and is soon changed."[41] All bodies contain a vital spirit "by which the operations of the soul are dispensed."[42] Even "the Excrements of the bodies of living Creatures, retain a portion of vitall spirits: and therefore we must not deny them life."[43] The fact that the sympathetic action of this spirit operates over great distances has been proven by the accounts of the weapon-salve.[44] Blood is an ingredient of this salve and we may conclude that it and "*sweat* are most stufft with *vital spirits*."[45]

Having established a theoretical basis for true medicine, Irvine next turned to practice. No one should be a slave to Galen, Avicenna, or Paracelsus. "These were great men, but when they gave themselves to contentious disputes to defend their own opinions, they much erred many times from the truth."[46]

Like other Paracelsians, Irving rejected phlebotomy[47] and the cauterization of wounds.[48] He did accept potable gold and claimed that he gave a plainer recipe for it than any other writer.[49] He relied heavily on Oswald Croll for his discussion of signatures and he insisted that medicines "bear the *signature* of the *disease*, and of the *part principally affected*."[50]

---

[38] *Ibid.*, p. 10 (Aphorism 70).

[39] *Ibid.*, p. 13 (Aphorism 94).

[40] *Ibid.*, p. 14.

[41] *Ibid.*, p. 22. For a discussion of similar cures and their relation to folk medicine, see Martha Baldwin, "Expanding the Therapeutic Canon: Learned Medicine Listens to Folk Medicine," a paper presented at Duke University, 18 April 1998.

[42] Irvine, *op. cit.*, p. 22.

[43] *Ibid.*, p. 25.

[44] *Ibid.*, pp. 30–31.

[45] *Ibid.*, p. 41.

[46] *Ibid.*, p. 56.

[47] *Ibid.*, p. 50.

[48] *Ibid.*, p. 56.

[49] *Ibid.*, p. 63.

[50] *Ibid.*, p. 64.

Above all, Irvine turned to transplantation as a painless cure. For him, "Transplantation is, when by means of a Magnetick, we put the Disease into a plant, or another living creature, the Patient being fully and wholly cured."[51] As an example he chose an infant with a high fever. Cucumbers were placed next to him "when he is asleep; the *Cucumbers* will wither, and the Child be cured. Some lay young *whelps* to the feet of young Children in their Cradles, and so often recovered."[52]

But this is a "magnetic" medicine and various "magnets" might be employed. Of special strength is *"dried mans flesh,"* which should be taken "from the body of a man that dies a *violent death*, and yet while it is warm."[53] Other magnets described included blood, dung, urine, sweat, hair, and the parings of the nails, spit and the "excrements of the nose."[54] Here Irvine digressed to teach his reader a secret.

> Let the *blood* with the most excellent parcell of the whole body, be joyned in a true proportion, by the best way possible, and put them into a *natural vessel*, well shut up, and set under a hen to hatch, and in the product thou wilt finde a thing performing many miracles, *coagulated* in the *shape of a man*.[55]

Here he reflects the homunculus of Paracelsus. The Paracelsian influence continues in an appendix in which he uses the Mumia of Paracelsus for transplantation.[56] In this case enough blood is taken from the right arm of a sick person to fill two egg shells, which are subsequently closed and placed under a hen for fourteen to twenty days. On opening "thou shalt find the *blood* of the sick-man by that *digestion* become monstrous, and of a most vile smell." In this form it is mixed with bread or meat and given to a pig or dog to eat.

> Then thou shalt perceive sensibly the Disease to leave the *man*, and infest the beast. . . . By this true *Mumia* of *Paracelsus*, may be cured *Leprosies, Jaundise, Gouts, Hydropsies, Consumptions, Cancers*, and other Diseases. . . . take notice that *putrifaction*, or rather *fermentation*, is here, the *Key* in the hand of Nature . . .[57]

[51] *Ibid.*, p. 68.
[52] *Ibid.*, p. 73.
[53] *Ibid.*, p. 78.
[54] *Ibid.*, pp. 82–89.
[55] *Ibid.*, p. 91.
[56] *Ibid.*, p. 97.
[57] *Ibid.*

OTTO TACHENIUS (FL. 1644–1699) AND HIPPOCRATIC CHEMISTRY

The late seventeenth century was a period of increasing interest in chemistry and Otto Tachenius is important in drawing additional attention to the acid-alkali theory. Of his life we know relatively little.[58] A German, born in Westphalia, he was reputedly still alive in Venice in 1699. At first he made a living as a pharmacist in a number of eastern German cities, but he went to Italy in 1644 and took an M.D. degree at Padua in 1652. His first works were attacks on one Helwig Dieterich, but his most important work was the *Hippocrates chimicus* that appeared first in 1666. This, with its *Clavis* and a frequently appended treatise on the viperine salt went through numerous editions in the next decade. Among the seventeenth-century printed books in the National Library of Medicine are twelve Latin editions produced between 1669 and 1678, as well as English translations of 1677 and 1690.

Tachenius was convinced that in ancient times the true natural philosophy had been developed by Hippocrates and that this was the philosophy of the chemists of his own day. As evidence he referred to the Hippocratic treatise, the *de Diaeta* (now referred to as the *Regimen*).

> Now all animals, including man, are composed of two things, different in power but working together in their use, namely, fire and water. Both together these are sufficient for one another and for everything else . . .
>
> . . . fire and water, as I have said, suffice for all things throughout the universe unto their maximum and the minimum alike.[59]

Although the greatest of Greek philosophers surely was Hippocrates, Galen was not to be despised as some would urge. Galen had travelled throughout Egypt and although chemistry then was a secret art, he had learned of it.

---

[58]On his life, see the short account by Marie Boas Hall in the *Dictionary of Scientific Biography*, vol. 13 (New York: Scribners, 1976), pp. 234–35. This includes a bibliography. Marie Boas [Hall] has also written a useful paper, "Acid and Alkali in Seventeenth Century Chemistry," *Archives Internationales d'Histoire des Sciences*, *9* (1956), 13–28. On the chemistry of Tachenius, see J. R. Partington, *A History of Chemistry*, *2* (London: Macmillan, 1961), 291–97. And on the Hippocratic background, see Ingo W. Müller, "Untersuchungen zum Hippokratesverständnis von Otto Tachenius und Michael Ettmüller," *Medizinische Journal*, *22* (1987), 327–341.

[59]Hippocrates, *Regimen*, trans. W. H. S. Jones, *The Loeb Classical Library*, *4* (London: Wm. Heineman, 1931), pp. 223–447 (231).

> *Thus he wrote of the wonderful efficacy of* Fire *and of* Treacle *to* Piso, *a piece most worthy to be read. And he highly extols* Aeschiros, *the Empirick, his Master, from whom he learned how to burn* Crabbs.
>
> *Wherefore* Galen's *Praise remains entire, and He will always be accounted a most sagacious Person; To whom, as well as to Hippocrates, a Reverence is due, as to the beginners and producers of many, if not all Good Things.*[60]

Turning to Aristotle, Tachenius stated that he had depended on three causes: material, formal, and efficient. Hippocrates had reduced these to two, fire and water. These two were properly understood to be acid and alkali "because all things in the Universe are made up of those two Universal principles." But to these "a Third doth inseparably adhere: Hence arose *Sal, Sulphur* and *Mercury* . . ."[61]

Tachenius fully accepted the macrocosm-microcosm analogy, which he employed in his explanation of chemical processes. Thus,

> . . . as the Ingenious *Cosmopolita* shews . . . as the Rain receives That Vertue of Life, and by the Sun Beams joyns it with the Alcaly of the Earth; so also the same Vertue of Life is attracted into the *Microcosme* by Inspiration, and is fixed by the Solar Beams of the Heart into the Alcaly, or Radical Moisture of Animals; . . .[62]

No less than van Helmont, Tachenius accepted the indestructibility of matter: "no Matter can be so destroyed, but it will remain under some Form or other."[63]

Understandably, the interest of Tachenius in acids and alkalis led to a special concern with the formation of salts.

> . . . the hungry Acid desires to be satisfied with its like, and *That* like (i.e.) food, it dissolves and is delighted with it; . . . For example, The Acidity of *Aqua Fortis* is as an empty Stomach, which desires to be satisfied: Silver being given it for food, it dissolves it, and is pleased with

---

[60]Otto Tachenius, *His Hippocrates Chymicus Discovering The Ancient foundations of the late Viperine Salt with his Clavis there unto annexed,* trans. I. W. (London: Nath. Crouch, 1677), sig. b3[v].

[61]Otto Tachenius, *His Clavis to the Ancient Hippocratical Physick or Medicine; Made by Manual Experience in the very Fountains of Nature. Whereby, Through Fire and Water, in a Method unheard of before, the Occult Mysteries of Nature and Art are Unlocked and clearly Explained by a Compendious way of Operation* (London: Printed by Tho. James and are to be sold by Nath. Crouch, 1677), p. 2.

[62]*Ibid.*, p. 82.

[63]*Ibid.*, p. 49.

> it; but when you cast in a *Physical Drug*... into this Solution; I mean *Copper*, with which, for the similitude between them, it is more delighted than with the *Silver*, presently it deserts the Silver, and again dissolves the Copper, and the whole Solution becomes green... Again, If you put *Iron* into *Aqua Fortis*... impregnated or loaded with Copper, in regard Copper is of harder Solution and Concoction than Iron, the Water presently leaves the Copper and dissolves the Iron.[64]

The digestive process seen in the action of acids on alkalis with the formation of salts is fundamentally the same in man since "It must needs be also in the Body of Man, since *Nature is in every thing alike*."[65] Thus we find that in the human stomach there is "a Volatile and never failing *Acid*; by the help of which, Aliments, which are full of Volatile Alcaly... are transmuted and turned into Chyle or Juice, first Sourish, and afterwards, by little and little into Salt (i.e.) mature."[66]

It seemed clear that the digestive process was simply another case of an acid-alkali reaction.

> ... it is the office of the Stomach to change one thing into another by the help of the two faculties (i.e.) the innate Acid, and the acquired Alcaly... So that the aforesayd Acid... assumes the Alcaly in the Aliments, and whatsoever is not repugnant to its own nature, it coagulates it with it self, and increaseth into a bulk which is vulgarly called *Nutrition*...[67]

Tachenius found palingenesis of particular interest. Here he referred to the older work of Duchesne in which the ashes of a rose were placed in a glass vessel that was then warmed by a candle.

> *and then that fine and impalpable Ashes of it self, would represent the evident form of a Rose, encreasing and growing by degrees, manifestly exhibiting the figure of a living florid Rose, which shadowy representation, when the Vessel was removed from the heat, would again return to ashes.*[68]

---

[64] *Ibid.*, p. 74.

[65] *Ibid.*, p. 75.

[66] Tachenius, *Hippocrates Chymicus*, p. 38.

[67] *Ibid.*, p. 49.

[68] *Ibid.*, p. 72. On palingenesis, see Jacques Marx, "Alchimie et Palingénésie, *Isis*, *62* (1971), 274–89; François Secret, "Palingenesis, Alchemy and Metempsychosis in Renaissance Medicine," *Ambix*, *26* (1979), 81–99; and Debus, *The French Paracelsians*, pp. 161–63.

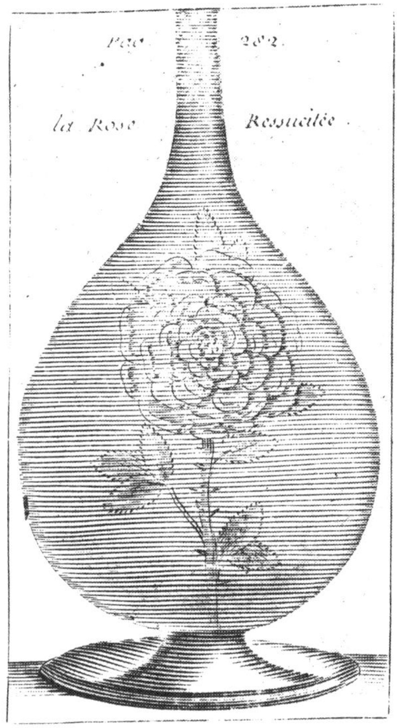

The resurrection of a rose from its ashes. From Pierre Le Lorrain, the Abbé de Vallemont, *Curiositez de la Nature et de l'Art sur la Vegetation: ou l'Agriculture, et le Jardinage* . . . (Paris, 1711). From the collection of the author.

Tachenius suggested that this process should be carried out with poppy seeds and dew, and he explained that the success of the experiment was due to the internal action of acid and alkali. Unfortunately he had been unable to carry this out himself "for want of fruitful and pure Dew" in Venice, but he had no doubt that it could be done.

> I add, that in the whole school of Philosophy, there is nothing, though never so diligently sought for, which doth more lively demonstrate to us the Mystery of the Resurrection, than these proper instruments of Nature, being cleansed from their Refuse, by an Hippocratist, and again conjoyned, re-united, and re-produced, if the said Mystery be at all demonstrable by Natural things.[69]

[69]Tachenius, *Hippocrates Chymicus*, p. 75.

For Tachenius chemistry was clearly the basis for all true understanding of natural philosophy.

> Our *Chymists* have directed themselves through the path of every Science, to enter into all Experience by Art, according to the course of Nature in her univocal Principles. For 'tis only Chymistry, which is the Glass of the true Understanding, shewing it how to feel and see Truths in a clear light, and therefore *Tabula Smaragdina* saith, By this kind of demonstration *all* Obscurity is banished and expelled from Man . . .[70] But why do I spend time in mentioning these things, there is not an Old Woman in Italy, but will inveigh against the opposers of this Art? for without It, it is impossible for them to find out any thing to Colour and Dye their Hair.[71]

ACID AND ALKALI: THE INITIAL REACTION

Otto Tachenius had been attacked as early as 1657 for his earliest publications and his promotion of the viperine salt. His opponent was Johannes Zwelfer (1618–1668), a pharmacist from the Rhenish Palatinate who–like Tachenius–had taken an M.D. at Padua.[72] Then returning to Vienna, Zwelfer wrote lengthy commentaries on the renowned *Pharmacopoeia Augustana.* While preparing a new commentary in the mid-sixties, Zwelfer read the *Hippocrates Chimicus*, to which he prepared a refutation exceeding two hundred pages in length. This is a detailed chemical discussion of the work, much longer than the original. In addition to attacking the viperine salt, Zwelfer complained that Tachenius did not satisfactorily define acids and alkalis. Dated 1667, a year before Zwelfer's death, it did not appear until 1672 as an appendix to the new edition of the *Pharmacopoeia Augustana Reformata.*[73]

---

[70] *Ibid.*, sig. a3^{v}.

[71] *Ibid.*, sig. b2^{v}.

[72] See Partington, *2*, pp. 296–97 and the biographical sketch by R. P. Multhauf in the *Dictionary of Scientific Biography*, vol. 14 (New York: Scribners, 1976), pp. 639–40.

[73] *Pharmacopoeia Augustana Reformata Cum ejus Mantissa & appendice, Simul Animadversionibus, Authore Joanne Zwelfer . . . Cui annexa est ejusdem Authoris Pharmacopoeia Regia Ut & Mantissa Spagyrica. Accessêre etiam huic Editione Bini Discursus Apologetici contra. Otth. Tachenium & Francisc. Verny* (Dordrecht: Apud Vincentium Carmax, 1672).

Robert Boyle turned his attention to acid-alkali theory in his *Experiments, Notes &c. About the Mechanical Origine or Production Of divers Qualities* (1675).[74] Here he included his "Reflections Upon the Hypothesis of Alkali and Acidum," but in contrast with others he did not discuss this concept in relation to medicine. For Boyle the concept was related to that of the three Paracelsian principles, salt, sulphur, and mercury, and he recognized the fact that its proponents treated them as elements, believing that acid and alkali are in all things.[75] But, Boyle noted,

> . . . it is not enough to say, that an acid, as such, performs these things, and an alkali so many others, that they divide the operations and phaenomena of nature . . . between them; since assertions of such great moment ought not to be advanced or received without sufficient proof.[76]

He believed that there were great differences between the variety of acids and alkalis and for that reason alone, Boyle remarked, "I am induced to question whether the acidum and alkali, we are speaking of, have the simplicity, that philosophy requires in principles."[77] Above all, philosophers require that definitions be clear,

> yet I do not much wonder, that the definitions given us of acidum and alkali should be but unaccurate and superficial, since I find not, that they have themselves any clear and determinate notion or sure marks, whereby to know them distinctly, without which chemists will scarce be able to form clear and settled notions of them.[78]

The list of observations given by these followers of acid and alkali such as the increase of heat on mixing, the formation of bubbles, or taste are not satis-

---

[74]Robert Boyle, *Experiments, Notes &c. About the Mechanical Origine or Production of divers Qualities: Among which is inserted a Discourse of the Imperfection of the Chymists' Doctrine of Qualities; Together with some Reflections upon the Hypothesis of Alcali and Acidum* in Boyle, *Works* (1772) vol. 4, pp. 284–92. ("Reflections Upon the Hypothesis of Alcali and Acidum"). See also Partington, *2*, p. 539 and Boas [Hall], *op. cit.*

[75]Boyle, "Alcali and Acidum," p. 284.

[76]*Ibid.*, p. 286.

[77]*Ibid.*

[78]*Ibid.*, p. 287.

factory for identification since seldom does one substance exhibit all of these properties.[79]

In the course of his discussion, Boyle rejected the traditional use of sympathy and antipathy as an explanation of the interaction of acids and alkalis. "For I look upon amity and enmity as affections of intelligent beings; and I have not yet found it explained by any, how those appetites can be placed in bodies inanimate and devoid of knowledge, or of so much as sense."[80] Ever the sceptic, Boyle concluded that those who hoped to "extend this narrow chemical doctrine to the whole object of natural philosophy, . . . must do more, than I expect they will be able before they can make me their proselyte."[81]

FRENCH OPINION

A French proponent of the acid-alkali theory was François de Saint-André (fl. 1670–1725) of the Faculty at Caen whose *Entretiens sur l'acide et l'alcali* appeared first in 1672. Presented in dialogue form, Saint-André praised the recent work of the anatomists who illuminated our understanding of the humors and their uses, and the chemists who had taken us from the darkness of the four elements.[82] Nevertheless, his purpose in writing was to deal with the chemical concepts being discovered and their value.[83] He turned first to the principles. The three active principles are the Paracelsian triad, salt, sulphur and mercury, while the passive principles are water and earth–a familiar late seventeenth-century presentation.[84] However, these principles were dependent on acids and alkalis. The active principles were composed of acid and alkali-salts, while water served as their dissolvent and earth acted as a bonding agent for them.[85]

---

[79] *Ibid.*

[80] *Ibid.*, p. 289.

[81] *Ibid.*

[82] Fran. Andre, Dr. in Physick, of the Faculty of Caën, *Chymical Disceptations: Or DISCOURSES upon ACID and ALKALI. Wherein are Examined the OBJECTIONS of Mr. Boyle against these PRINCIPLES. Together with a REPLY to a Letter of Mr. S: Doctor of Physick, & Fellow of the Colledg of ***, wherein many Errors are corrected, touching the Nature of these two SALTS, Faithfully rendred out of French into English by J. W. To which is added, by the Translator, a Discourse of Phlebotomy, shewing, the Absolute Evils, together with the Accidental Benefits thereof, in some cases* (London: Printed by Tho. Dawks and Benj. Allport, 1689), p. 3.

[83] *Ibid.*, p. 4.

[84] *Ibid.*, pp. 5–6.

[85] *Ibid.*, pp. 6–10.

Saint-André was a corpuscularian who believed that all bodies in nature were composed of atoms or particles. Shape played a significant role in their differentiation. Thus, acid salts were evident by their taste, smell, and the fermentation they produce with alkalis.[86] There seemed little doubt that they were composed of small sharp-pointed particles that acted on the porous alkali particles.[87] The existence of acid-alkali color indicators, clearly noted as early as 1632 by Edward Jorden and developed by Robert Boyle in succeeding decades, made the presence of acids and alkalis undeniable.[88]

But if a principle system based upon acid-alkali theory defined matter, chemical action proceeded through fermentation and

> By the Word Fermentation, I mean an Internal Motion of all the parts of Bodies which are fermented in such Manner that they take no more the same place nor scituation as they had before, and that they change consequently, or at least alter very much the nature of the bodies which are fermented.[89]

Saint-André was well aware that there were different kinds of fermentation, some of which occurred with an effervescence and others which did not. Here he gave as examples an egg hatched by a hen and a chemical mixture in which the evidence of change is noted by an increased temperature.[90]

In the second edition of his work (1677), Saint-André added a reply to Boyle's objections. Here he noted that Boyle believed that it had not been proven that acid and alkali are present in all bodies, and that the division of salts into acid and alkali was purely arbitrary.[91] Saint-André rejected both statements, arguing that there were truly great differences between acids and alkalis based upon their fundamental identifying shapes.[92] He then went on to discuss Boyle's observation of the different effects of the action of acids on

---

[86] *Ibid.*, p. 11.

[87] *Ibid.*

[88] *Ibid.*, p. 65. On Edward Jorden's use of color indicators, see Allen G. Debus, "Edward Jorden and the Fermentation of Metals: An Iatrochemical Study of Terrestrial Phenomena." In *Toward a History of Geology: Proceedings of the New Hampshire Interdisciplinary Conference on the History of Geology. September 7–12, 1967*, ed. Cecil J. Schneer (Cambridge, Mass.: M.I.T. Press, 1969), pp. 100–121. Boyle's use of color indicators is discussed by Marie Boas in *Robert Boyle and Seventeenth-Century Chemistry* (Cambridge: Cambridge University Press, 1958), pp. 133–38.

[89] *Ibid.*, p. 76.

[90] *Ibid.*, p. 77.

[91] *Ibid.*, pp. 86–100.

[92] *Ibid.*, pp. 101–3.

alkalis. This, Saint-André was convinced, resulted from the different natures of specific acids and alkalis.[93] And finally, Boyle argued that acid-alkali theory could not explain "manifest qualities" such as gravity, light, or occult qualities as one finds in magnetism. Not so, replied Saint-André, who described magnetic effects in terms of acid particles leaving and returning from a loadstone.[94]

Finally, in a "Reply to a letter of Mr. S. touching the nature of Acid and Alkali," Saint-André turned to the subject of digestion. Noting that both Sylvius and de Graaf attributed the fermentation of the chyle to the acidity of the pancreatic juice, he stated that it was well known that this juice was not acid.[95] Rather, the chyle had an acid salt taste and it was known that acid-salt bodies were a mixture of acid and alkali. From this it would be concluded that the chyle was an acid salt and that the bile was a volatile alkali.[96] The two were fermented as soon as they were dissolved in the pancreatic juice. Although Saint-André did not mention Tachenius, we can see here an extension of this concept. And if Boyle did not discuss any medical aspects of this theory, Saint-André turned to the digestive process as an important example.

In 1683, a Dr. F. Bertrand of the Marseilles College of Medicine published his *Reflexions nouvelles sur l'Acide et sur l'Alcali.* Here he sought to demonstrate that neither acid nor alkali were the principles of mixed bodies. He was particularly concerned with an examination of the work of Tachenius because of the number of people he had known who were convinced of the truth of his work.[97] Bertrand sought to show that the views of Tachenius did not correspond with those of Hippocrates.

He was concerned first with definitions. There were some who said that an acid is that which ferments with an alkali and that an alkali is that which absorbs an acid. Bertrand sought rather to present a corpuscular definition. "I say that acid is a liquid body composed of small particles that are stiff and pointed, a little like very delicate and fine needles."[98] On the other hand,

[93] *Ibid.*, pp. 102–3.

[94] *Ibid.*, p. 120.

[95] *Ibid.*, p. 141.

[96] *Ibid.*, p. 143.

[97] Mr. Bertrand, Docteur en Medicine Aggregé au College des Medecins, de Marseille, *Reflexions nouvelles sur l'Acide et sur l'Alcali: Où apres avoir demontré que ces deux sels ne peuvent pas être les principes des Mixtes, On fait voir le veritable usage qu'on en peut faire dans la Physique & dans la Medecine* (Lyon: Thomas Amaulry, 1683), sig. ã7′–ã8′.

[98] *Ibid.*, p. 3.

alkali is a solid earthy body with pores of different structure. Solution of an alkali by an acid consists of the taking away of the acid points.[99]

Bertrand recognized that van Helmont was among the first to begin to understand the properties of acids and alkalis[100] and he was well aware of the debate between Tachenius and Zwelfer.[101] However, he felt that some had gone too far in claiming that their system of acid and alkali extended throughout nature.[102] It may be that fermentation accompanies the generation of plants and that the seeds cannot develop without an acid that is proper to excite this fermentation, but can we go further? What agitates the different juices of the plant and serves to make it grow? What causes it to grow in one way and not another? And what gives flowers and fruits their specific colors? We cannot say that this is all due to acid and alkali.[103]

Moreover, what is the cause of the heat of the blood and its continual movement? Bertrand believed that the agitation of the blood was due to subtile matter that gives motion to the larger parts.[104] Here he referred to Descartes and his spherical particles.[105] And although these new chemists argued that acid and alkali extended vast distances into the heavens,[106] and that they were the principles of all natural things, it was impossible to extract them from many substances.[107]

In the *Hippocrates chimicus*, Tachenius claimed that all sublunary things are composed of acid and alkali.[108] Here he had turned to Hippocrates as his authority and asserted that only chemists can understand these mysteries. But Hippocrates had referred to fire and water, Bertrand noted, and Tachenius erred in assuming that they could be equated with acid and alkali.[109] Indeed, for Tachenius acid and alkali should be the basis for all things, animal, vegetable, and mineral. But Bertrand found it surprising to see Tachenius put forth passages from Hippocrates that really had no connection with the sub-

[99] *Ibid.*, pp. 4–5; see also p. 93.

[100] *Ibid.*, p. 30.

[101] *Ibid.*, p. 8.

[102] *Ibid.*, p. 42.

[103] *Ibid.*, p. 46.

[104] *Ibid.*

[105] *Ibid.*, p. 50.

[106] *Ibid.*, p. 57.

[107] *Ibid.*, p. 72.

[108] *Ibid.*, p. 108.

[109] *Ibid.*, p. 112.

ject. This Tachenius explained by arguing that Hippocrates sought to present his work as a mystery so that only the worthy would understand.[110] This was a typically alchemical style of presentation.

Actually, the acid and alkali in the human body were not principles, Bertrand argued, but rather spirituous and oily parts of the blood placed in the body for the necessity of their structure and function.[111] And as for Tachenius's interest in palingenesis, the artificial resuscitation of plants, he had simply borrowed this from Athanasius Kircher's *Mundus Subterraneus.* In any case, one could not argue as Tachenius did that palingenesis gives proof of resurrection.[112]

Having demolished the theory of acid and alkali to his own satisfaction, Bertrand proceeded to the second section of his book in which he took up subjects such as inflammability, earthquakes, thunder, lightning, effervescence, the cause of color, and the generation of minerals.[113] He constantly referred to the importance of chemistry for the physician since this would make evident the nature of the humors and indicate the means of reducing them into a true temperament when sickness had altered them.[114] And we would surely understand more of the blood with an application of new instruments. Thus Mr. Landoüillete had studied the blood with his microscope and found it to be composed of small red balls in a crystalline liquor. Each of these balls is composed of six smaller ones and they have a characteristic motion from the bottom to the top of the vessel.[115]

Finally, Bertrand discussed diseases in terms of acid and alkali, blaming them for malign and contagious illness.[116] But too many chemists had sought a universal medicine. "I doubt myself if this universal medicine exists, and whatever one says of the alkahest of Paracelsus and van Helmont, I scarcely believe that they ever found a remedy filled with the virtues they attribute to it."[117]

The views of Tachenius on the ancient Greek origins of the acid-alkali hypothesis found a supporter in François Aignan (1644–1709), a physician to the King and a doctor of the faculty of Padua. In his *L'Ancienne Medecine a la*

---

[110]*Ibid.*, p. 136.
[111]*Ibid.*, p. 152.
[112]*Ibid.*, p. 159.
[113]*Ibid.*, pp. 166–67.
[114]*Ibid.*, p. 169.
[115]*Ibid.*, p. 172.
[116]*Ibid.*, pp. 222–23.
[117]*Ibid.*, pp. 309–10.

Mode, ou le Sentiment Uniforme d'Hippocrate & de Galien sur les Acides & les Alkalis (1693), Aignan again turned to Hippocrates as the founder of the acid-alkali theory. Those who thought that this was a new medical system simply did not understand the ancients.[118] Not only was there evidence to be found in Hippocrates *de Diaeta (Regimen)*, but also in his *On the Ancient Medicine*, in which he referred to acids, salts, and other chemical terms.[119] In short, Aignan believed that Hippocrates was a chemist.

Aignan turned to Hippocrates as an authority in stating that the degeneration and transmutation of salts changed the natural state of the four humors.[120] But above all, Aignan turned to digestion as a chemical process due to a vital acid and a natural ferment that "is a corrosive vitriol and coagulant that *Hippocrates* knew well."[121] This acid coagulates milk and is the cause of ulcers, rheumatism, gout, and fevers.[122] Still, it is a necessity and brings about a transmutation at the end of the duodenum. "It is there that the food receives the seal and the character of a volatile salt which is its natural balm."[123] There is little doubt then that this acid has two characteristics, vital and vicious–the one necessary for life, the other a cause of disease.[124]

Aignan underscored another belief of the chemists three years later in *Le Prestre Medecine (The Priest Physician)*, in which he returned to the godly office of the medical doctor. For him medicine was an ecclesiastical institution and he listed ancient and modern men of the clergy who had also studied or practiced medicine.[125] Indeed, both the cure of the soul and the body required the support of God. It seemed only right then to accept the fact that cure by miracles and cure by pharmaceutical remedies were a part of the same natural science.[126]

In his search for medical texts to support his position, Aignan cited Hippocrates, Plato, and other ancients as well as Biblical figures. But Paracelsus

[118]François Aignan, Medecin du Roy, & Docteur de la Faculté de Padouë, *L'Ancienne Medecine a la Mode, ou le Sentement Uniforme d'Hippocrate & de Galien sur les Acides & les Alkalis* (Paris: Laurent d'Houry, 1693), pp. 80–82.

[119]*Ibid.*, p. 85.

[120]*Ibid.*, p. 92.

[121]*Ibid.*, pp. 102–4.

[122]*Ibid.*, pp. 104–6.

[123]*Ibid.*, pp. 134–35.

[124]*Ibid.*, p. 139.

[125]François Aignan, Medecin du Roy, & de M. le Cardinal de Furstemberg & Docteur de la Faculté de Padoué, *Le Prestre Medecin ou Discours Physique Sur l'établissement de la Medecine. Avec un Traite du Caffé & du Thé de France, selon de Systeme d'Hippocrate* (Paris: Laurent D'Houry, 1696), c. 1.

[126]*Ibid.*, c. 2, p. 76.

had placed his finger on this essential truth, "that to be a good physician, it is necessary to be a good theologian; medicine and theology are inseparable, so that one is forced to assert that these two professions should never be permitted to separate."[127]

Again Aignan turned to acid-alkali explanations in his description of fermentation, which he treated morally as well as physically. "I say once more that without this mystery of fermentation it is impossible to accomplish any physical operation that is worthy of a philosopher."[128] Indeed, fermentation makes possible all vegetation, transmutation, and the multiplication of species.[129] And again, Aignan drew on Paracelsian sources when he insisted that "God has placed in each country, and in each province, all that is necessary for medicine without having recourse to help from foreign countries– and this according to the belief of Hippocrates."[130]

ACID AND ALKALI: THE PROPER CAUSE OF DISEASE

In England a new debate over the cause of disease resulted from a 1696 publication by John Colbatch (d. 1729) titled *A Physico-Medical Essay Concerning Alkaly and Acid, so far as they have relation to the cause or cure of distempers, wherein is endeavoured to be proved, that acids are not (as is generally and erroneously suppos'd) the Cause of all or most Distempers, but that Alkalies are. . . .* Two years later he added *Some Farther Considerations Concerning Alkaly and Acid* (1698).

Colbatch was trained first as an apothecary and he was accepted as a licentiate of the Royal College of Physicians in 1696 and knighted twenty years later. He was a prolific author and his collected works went through several editions.[131]

Although Colbatch received considerable recognition during his lifetime, he had had relatively little formal education.

[127] *Ibid.*, c. 11.

[128] *Ibid.*, c. 6, p. 167.

[129] *Ibid.*

[130] *Ibid.*, c. 10, p. 215.

[131] On Colbatch, see Harold J. Cook, "Sir John Colbatch and Augustan Medicine: Experimentalism, Character and Entrepreneuralism," *Annals of Science*, *47* (1990), 475–505; Harold J. Cook, *The Decline of the Old Medical Regime in Stuart London* (Ithaca and London: Cornell University Press, 1986), p. 214.

> I am not ashamed to own that I have not had a University Education, which I account a Misfortune; but yet I have spent as much Mony to qualify me for the Practice of Physic, as most Men who have spent their times in the Universities, I have had ocular Demonstration for my Mony, when as some of our University-Men have had only the fusty Notions of old Aristotle for theirs. I have aim'd at Certainty in the Cure of Diseases, and if I can but once arrive to that . . . I care for no more.[132]

In his *Physico-Medical Essay* of 1696 Colbatch sought to refute those who argued that the cause of disease derived from acids.[133] Here he complained that there was much confusion regarding the identification of acids and alkalis. The true answer here was to be found in chemistry: "I own myself a profest Chymist, and in many things, tho not all, a Disciple of *Helmont*, and know nothing so delightful to me as Chymical Operations."[134] And yet, "I don't pretend to be so great, nor deserve as well of Mankind as Helmont, Paracelsus, &c."[135] Even that universal solvent, the alkahest, seemed to be a real possibility:

> The Alkahest, that universal Dissolvent of Helmont and Paracelsus, is a thing that has mad a great Noise in the World; and tho many have bin at vast charge in quest after it, yet I have scarce met with any (except one) in our time, who has been Master of it. But on some late Experiments made upon the Pyrittis, I am perfectly satisfied that the said noble Menstruum, is with little Trouble and charge to be obtained from this so common, yet most valuable Subject. In convenient season I may give an account of the Experiments I have made upon it.[136]

---

[132]John Colbatch, *Some Farther Considerations Concerning Alkaly and Acid . . . in A Collection of Tracts Chirurgical and Medical* . . . (2nd Corrected edition, London: Dan. Brown, 1704), p. 243.

[133]John Colbatch, *A Physico-Medical Essay Concerning Alkaly and Acid, . . . Wherein is endeavored to be prov'd, that Acids are not (as is generally and erroneously suppos'd) the Cause of all or most Distempers, but that Alkalis are* in *A Collection of Tracts*, p. 85.

[134]*Ibid.*, p. 107.

[135]*Ibid.*, p. 95.

[136]Colbatch, *Farther Considerations*, p. 245.

Colbatch's chemical interests had been fostered initially through his training in pharmacy, "which is the Foundation on which the Practice of Physic is built."[137] Not only must the pharmacist know the vegetable kingdom and the nature of animal bodies, he "must well understand the nature and operation of minerals and metals," which must "be learned by much reading and hard labor in the fire."[138]

Turning to specific diseases, Colbatch pointed to the significance of the acid-alkali color indicators. It was known that acids turn syrup of Violets red, while alkalis turn it green. This test proved those wrong who argued that an acidic blood caused disease, since "the *Serum* of the *Blood* changes the blue Color into a green one; therefore it cannot abound with *Acids*."[139] This general conclusion was borne out in the case of specific diseases, particularly scurvy, which is cured by the juices of lemons and limes which are natural acids.[140] It could be concluded from this that scurvy is due to alkaline particles in the blood.[141] An analysis of gout, rheumatism, and other diseases would confirm this result and "it is plain that Acids are not the Cause of Diseases, but Preservatives against them; and whatever is a Preservative from Distempers by parity of reason must be of use to repel them."[142] Such reasoning could be pushed a step further. Assuming that "Sulphurs are the active Principle in both Animal and Vegetable Juices; Alkalies in breaking and dividing the Particles excite Fermentations, and by degrees cause a total dissipation of them."[143] On the other hand,

> Acids are a sort of *Vinculum* to Sulphurs, and therefore suppress Fermentations: From whence I infer that a true Oleosum must be the great Medicin: yet it is not such a one as commonly goes under that name, . . . but it must be an *Acido-olesum*, an exalted Acid, by which the Life of Man is chiefly supported.[144]

[137]Colbatch, *Physico-Medical Essay*, p. 91.

[138]*Ibid.*, pp. 92–3.

[139]Colbatch, *Farther Considerations*, pp. 249–51.

[140]Colbatch, *Physico-Medical Essay*, pp. 119–21.

[141]*Ibid.*

[142]*Ibid.*, p. 150.

[143]Colbatch, *Farther Considerations*, p. 448.

[144]*Ibid.*, p. 449.

Although clearly influenced by the late seventeenth-century obsession with acid-alkali theory, Colbatch was also well aware that the hypothesis did not answer all problems related to natural phenomena.

> I have all this while bin talking as a Physician, but not as a Naturalist; for tho I account for the Cause and Cure of Diseases from *Alkaly* and *Acid*, yet I don't pretend to account for all the *Phenomena* of Nature from those two Principles, tho there have bin those who have undertaken so to do, but I conceive have bin very much mistaken. There are many *Phenomena* in Natural Bodies not to be accounted for by the mutual Conflict of those two Principles . . . such as *Motion* and *Rest*, *Light* and *Darkness*, *Magnetism*, or the Operation of the *Loadstone*, and a great many others, too many here to insert.[145]

It is clear that, for Colbatch, chemistry was to be the salvation of medicine. To further this belief he recommended that Parliament establish a chemical laboratory for research.

> If we could have a Public Laboratory at the Nations Expence, I shall hope to see that matter sooner effected than it is possible to be at the private Charge of any one Man. It costs the Nation some Hundred Thousands of Pounds a Year in buying Powder and Ball to destroy Mankind (tho in carrying on the most just War that ever was on foot) and why should there not be one Thousand Pounds a Year laid out in carrying on those Experiments that would tend to the saving of Mens Lives?[146]

The *Physico-Medical Essay* of Colbatch was answered by a surprising number of physicians. Richard Boulton (fl. 1697–1724), educated at Brasenose College, Oxford, and known best for his four-volume epitome of the works of Robert Boyle (1700), prepared *An Examination of Mr. John Colbatch His Books* (1698). Here he noted that "rational physicians" had rejected the acid-alkali hypothesis. Above all, since Robert Boyle had shown its fallacy, it was a wonder that anyone should advance it again.[147]

---

[145] *Ibid.*, pp. 260–61.

[146] *Ibid.*, p. 243.

[147] Richard Boulton, of Brazen-Nose College in Oxford, *An Examination of Mr. Colbatch His Books* . . . (London: A. & J. Churchill, 1698), sigs. $A5^{r\&v}$.

Also in 1698 Francis Tuthill published *A Vindication of Some Objections lately raised against Dr. John Colbatch his Hypothesis* and Charles Leigh, M.D., like Boulton a graduate of Brasenose College, wrote against Colbatch's cure of viper bite by acids. An otherwise unknown S.W. refuted the views of Colbatch on gout.[148]

We may treat the work of Thomas Emes and William Coward in more detail. Emes (d. 1707) was known as "The Prophet" and connected in some way with the Camisards in France. By many he was considered to be a quack and he prophesied that he would arise from the dead on 15 May 1708, which raised the hopes of his followers until that date.[149]

However, Emes also styled himself a "Chirurgo-Medicus" and he strongly opposed the work of Colbatch. In 1698 he prepared *A Dialogue Between Alkali and Acid* aimed directly at Colbatch. Written in a common dialogue form, it presented a discussion between Mr. Acid and Mr. Alkali. Opening with their meeting, Mr. Alkali greeted his opponent:

> Well met Mr. *Acid*, whither are you hurrying so fast, to some Heroe run through the Lungs, or the Heart?

To which Mr. Acid replied,

> I should hardly stay to tell you Mr. *Alkali*, but that I am engag'd to oppose you where-ever we meet, you *Principle of Death and Corruption*, I am always provok'd by you, you have done so much mischief in the World.[150]

The tone of the discussion is well set by this beginning. Of course, Mr. Acid represents the views of Colbatch whom Mr. Alkali refutes at every turn. Addressing Mr. Acid, he states that "I cannot but look upon you as a person of great Impudence, to accuse all the World of Learned PHYSITIANS as

---

[148]Francis Tuthill, *A Vindication of Some Objections lately raised against Dr. John Colbatch his Hypothesis* (1698); Charles Leigh, M.D., *A Reply to John Colbatch, upon his late Piece, concerning the curing the biting of a viper by acids* (1698): S.W., *An Examination of a late Treatise of the Gout: wherein John Colbatch's demonstration are briefly refuted* (1698).

[149]See the sketch by Gordon Goodwin in the *Dictionary of National Biography* (1921–1922), *6*, pp. 772–73.

[150]T. E. [Thomas Emes] , Chirurgo-Medicus, *A Dialogue Between Alkali and Acid: Containing Divers Philosophical and Medicinal Considerations wherein A late Pretended New Hypothesis, asserting Alkali the Cause, and Acid the Cure of all Diseases; is proved Groundless and Dangerous* . . . (London: R. Cumberland, 1698), p. 1.

Fools, or what is worse, as Knaves . . ."[151] And indeed, although Acid brags of his originality, Alkali notes that "your Discourses smell pretty strong of Dr. *Andre*, of *Acid* and *Alkali* Published in English in 1689."[152] In the course of the dialogue, the meaning of the testing by syrup of violets is discussed as well as the perpetual interest in the possibility of transmutation and the Helmontian alkahest.

Two years later, an anonymous *Letter to a Physician concerning Acid and Alkali* was published. This prompted Emes to publish a second attack on Colbatch entitled *A Letter to a Gentleman Concerning Alkali Being an Answer to a late Piece Intituled, A Letter to a Physician concerning Acid and Alkali* (1700).[153] Here in a lengthy prose presentation Emes repeated the points he had made earlier in his *Dialogue.*

William Coward (1657?–1725) had been a scholar at Wadham College and then became a fellow at Merton.[154] In contrast to Colbatch, he had earned a series of academic degrees (B.A., 1677; M.A., 1683; M.B., 1685; M.D., 1687). After practicing in Northampton, he moved to London in 1693. Interested in theology, he published on the subject of the soul in books that were burned because of his heretical views. Also an accomplished classicist, Coward translated the work of Dryden from English into Latin.

In his *Alcali Vindicatum: Or the Acid Opiniator not Guilty of Truth* (1698), Coward attacked the *Physico-Medical Essay* of Colbatch. He was certainly not an enemy of chemistry, but whereas Colbatch had openly acknowledged that he was a disciple of van Helmont, we find Coward referring more frequently to Boyle. He attributed the many errors he found in Colbatch to his lack of a university education.[155] Coward thought that a physician needed a thorough grounding in philosophy as well as experience and he favored a full twelve years' training in British universities.[156]

Colbatch had referred to scurvy, gout, and rheumatism in support of his views. Again, the case of scurvy may be singled out. Coward was willing to

---

[151]*Ibid.*, p. 5.

[152]*Ibid.*, p. 9.

[153]Thomas Emes, *A Letter to a Gentleman Concerning Alkali, Being an Answer to a late Piece Intituled, A Letter to a Physician concerning Acid and Alkali* (London: Thomas Speed, 1700).

[154]See the sketch by Leslie Stephen in the *Dictionary of National Biography* (1921–1922), vol. 4, pp. 1298–1299.

[155]William Coward, M.D. of the Colledge of Physicians, London, *Alkali Vindicatum: Or The Acid Opiniator not Guilty of Truth. Being an Impartial Enquiry into the Fallacious Reasons and Erroneous Philosophy of a late Physico-Medical Essay Touching Alcali & Acid* (London: Tim. Childe, 1698), p. 120.

[156]*Ibid.*, pp. 123–24.

admit that the "Blood of Scorbutick Persons hath by much a greater quantity of Volatile Alkalious Particles, than that of sound Persons" and that after long voyages at sea "Seamen and Mariners have been cured of the Scurvy by eating Oranges and Lemons in Spain, after their return."[157] On the surface this might support Colbatch, but Coward argued that in reality the blood is "impregnated with a Marine Salt, not a fixed Alkali," and although oranges and lemons occasionally cured scurvy, the true cause of cure is that a purging regimen had removed the Scorbutick matter. Therefore "you ought not to make a general conclusion . . . That Acids are the only proper Medicines in the Scurvy."[158]

> . . . I say upon the whole, the Method of Alcalies is certainly more safe and prudent. A method by *Acids* may possibly be sometimes safe; I say possibly, because, some Patients cannot be killed by any irregular Practise whatever, so it be not down-right Poison[159]

Concern with the chemical action of physiological processes led Coward to an interest in fermentation. In his *De Fermento* (1705) he turned to digestion to argue that heat resulted in the "transmutation" of food.[160] Well versed in the current literature, Coward relied heavily on Willis's *De fermentatione*, but also turned to Glisson, Mayow, and other recent authors in his discussion of the digestive process.[161]

ACIDS, ALKALIS, AND THE JOURNALS

The French works by Saint-Andre, Bertrand, and Aignan, and the English debate over the views of Colbatch, represent only a small sampling of the late seventeenth- and early eighteenth-century interest in acids, alkalis, and fer-

[157] *Ibid.*, p. 57.

[158] *Ibid.*, p. 130.

[159] *Ibid.*, p. 53.

[160] William Coward, M.D. & Collegii Merton Oxon. Socio, *De Fermento Volatili Nutritio Conjectura Rationalis. Quà ostenditur Spiritum Volatilem Oleosum é Sanguine suffusum, Esse Verum a Genuinum Concoctionis ac Nutritionis Instrumentum* (London: J. H. pro T. Bennet, 1705). Here the title page includes the following quotation from van Helmont: "Non est calor, sed est alia facultas quaedam, quae verè atque formaliter transmutat Alimenta, Eámque Fermentorum nomine Designavi." Cook discusses the work of Coward in "Sir John Colbatch," 495.

[161] *Ibid.*, pp. 3, 16, 27, *passim.*

mentation. If we turn to some of the book reviews that appeared in the learned journals of the period, we note a continuing interest in chemical interpretations of physiology stemming from the seminal works of van Helmont, Sylvius, and Tachenius. Most of the chemical and medical works reviewed mention the acid-alkali theory and fermentation. Here we may well turn to two journals that specialized in reviews, the *Journal des Sçavans* and *The History of the Works of the Learned.*

In 1687 a review of Claude Foüet's *Nouveau Systéme de bains et eaux minerales de Vichy, fondé sur plusiers experiences & sur la doctrine de l'Acide et de Alcaly* (1686) appeared in the *Journal.*[162] Foüet commented on the rapid progress of the acid-alkali theory, giving full credit to Tachenius even though he had only pretended that the true originator had been Hippocrates. The reviewer noted that the theory had been opposed by Bertrand in his *Reflexions nouvelles sur l'Acide et sur l'Alkali,* which in turn was refuted by Foüet. In the same volume the *Tractatus duo* (1687) of Raymond Vieussens was reviewed. Here the second treatise dealt specifically with the nature, causes, and conditions of fermentation.[163] We will discuss his work at length in the next chapter.

Both the *Journal* and the *History* reviewed the *Dissertationes medicae* (1701) of Archibald Pitcairne, another author we will turn to later.[164] In the *Journal* it is noted that Pitcairne had rejected the acid-alkali theory, while the reviewer in the *History* noted that Pitcairne is "of the Opinion, that the *Acids* and *Alkalis* do neither good nor bad, that they neither nourish nor cure any Disease . . ."[165]

The first volume of Francis Bayle's *Institutiones Physicae* (1700) was reviewed the following year in the *History of the Works of the Learned.* Here it was noted that Bayle "applys himself particularly to explain the Nature and Effects of *Acids* and *Alkali,* the knowledge of which is so useful in Physick," and that he was especially interested in the nature, effects, and causes of fermentation, which he was convinced was based on acids and alkalis.[166]

---

[162] *Journal des Sçavans* (Amsterdam: Wolfgang, Waesberge, Boom & van Someren, 1687), p. 222.

[163] *Ibid.,* p. 509–15.

[164] *Journal des Sçavans, 30* (1702) (Amsterdam: Waesberge, Boom & Goethals, p. 1703), p. 332; *The History of the Works of the Learned: or An Impartial Account of Books Lately Printed in All Parts of Europe, With a Particular Relation of the State of Learning in Each Country* (London: H. Rhodes, J. Harris, T. Bennet, A. Bell, D. Midwinter, and T. Leigh), *4* (1702), pp. 145–50.

[165] *Ibid.,* p. 150.

[166] *History, 3* (1701), pp. 143–47 (146–47).

Saint-Andre's *Reflexions sur la Nature des Remedes* (1701) was reviewed in the *Journal* in 1704. Here the reviewer noted that because Hippocrates' *On the Ancient Medicine* referred to acids, alkalis, and other chemicals, Saint-Andre was convinced that he had had a perfect knowledge of chemistry.[167] This was an opinion firmly rejected by Gerard Goris in *Medicina contempta propter λογμαχιαν vel Ignorantiam Medicorum* (1700), which was reviewed in the *History* in the same year. Here Goris spoke out against those innovators who had stolen from the ancients and he attacked the Paracelsians and all those whose medicine was solely chemical or who devoted themselves entirely to the search for the Philosophers' Stone. Rather, he felt that Galenical and chemical medicine should be united. It is interesting that the reviewer noted that Goris "is of the Opinion that the Knowledge of Mathematicks is of no use in the Practice of Physick, and looks upon what some Modern Authors have taught of the Insensible Parts and their Mechanical Action to be of no Use."[168]

An anonymous *Explication of some Phaenomena which depend on Fermentation* was reviewed in the *History* in 1702. Normally an explanatory device of the chemists, a different approach was emphasized by this author who had an interest in the role of motion in fermentation processes. Noting rarefaction of soft bodies, changes in texture and temperature, he concluded that "we cannot deny that Motion is the cause of all these changes."[169] Turning specifically to the fermentation of grapes, he cited Descartes, who spoke of ethical matter insinuating itself into the smallest parts and causing motion "capable to turn upside down the most solid Bodies, and to divide the Texture of those that are most firmly united."[170] In short, this author believed that fermentation was due to the vortical motion of the ether.

In contrast to the Cartesian discussion of fermentation in the *Explication*, the *Journal* reviewed a traditional chemical work by Henri Louis de Rouviére in 1708, the *Reflexions sur la Fermentation et sur la Nature du Feu* (1708). Here Rouviére discussed the various types of fermentation. He rejected the atomic system of Democritus and he criticized Descartes while adhering to the five

---

167 *Journal, 32* (1704) (Amsterdam: Waesberge, Boom & Goethals, 1705), pp. 850–55 (853).

168 *History, 2* (1700), pp. 330–336 (335).

169 *History, 4* (1702), pp. 587–92 (588).

170 *Ibid.*, p. 589.

principle-element theory of matter favored by late seventeenth-century chemists.[171]

Not only did chemical explanations continue to claim the allegiance of a large number of physicians, but the works of van Helmont also continued to be read. In the 1704 volume of the *Journal*, there was a review of Matthew Heer Lusati's *Introductio in archivum Archei vitale & fermentale viri magnifici Joannis Baptistae Van-Helmont* (1703). Here the reviewer discussed the significance of the Helmontian archeus as the principle of growth and its necessity for the multistage digestive process.[172] And in the *Journal* for 1706 there was a review of Jean le Pelletier's work on van Helmont's *Alkahest ou le Dissolvant universal* (1706). Although le Pelletier evidently did not know how to make the alkahest himself, he believed that others had known the secret. Above all, he recommended to the reader George Starkey's alchemical classic, *Secrets Revealed*.[173]

In the volume for 1707 there was a review of a new edition (1701) of Charles Musitano's earlier (1683) work on chemical medicine.[174] Beginning with a history of chemistry which he traced back to Tubal-Cain, Musitano went on to a discussion of disease openly adhering to the doctrine of van Helmont. He, too, emphasized the archeus in maintaining health. His section on pharmacy divided remedies into two categories, those obtained from plants and those obtained from minerals. In both cases chemical preparations were essential.

The following year (1708) witnessed a review of Michael Bernard Valentini's new edition of the *Opera omnia* of van Helmont (1707). The reviewer centered his remarks on Valentini's introduction. He began with a discussion of the difficulties faced by those who had worked to reform the sciences, referring to Francis Bacon and Descartes before describing the difficulties faced by van Helmont. Indeed, the struggle of van Helmont against the Galenists is compared with that of Luther and Zwingli against the Roman Catholic Church.[175] A great achievement of van Helmont was to discover the errors of traditional medicine and to have based his findings on experience.

---

[171] *Journal*, *39* (1708) (Amsterdam: Janssons à Waesberge, 1708–1709), pp. 317–23 (318).

[172] *Journal*, *32* (1704) (Amsterdam: Waesberge, Boom & Goethals, 1705), pp. 47–48.

[173] *Journal*, *34* (1707) (Amsterdam: Janssons à Waesberge, 1707), pp. 831–38 (836).

[174] *Journal*, *37* (1707) (Amsterdam: Janssons à Waesberge, 1707), pp. 23–28.

[175] *Journal*, *41* (1708) (Amsterdam: Janssons à Waesberge, 1709), pp. 123–129 (124–25).

Valentini gave credit to Celsus, Fernel, and Paracelsus, but the most reliable author certainly was van Helmont.

CONCLUSION

The closing decades of the seventeenth century and the beginning of the new century witnessed a continuing interest in the chemical philosophy among physicians. In this sense medicine was different from the physical sciences, in which the mechanical philosophy would seem to have eliminated the chemical philosophy as a serious competitor by this time–at least among those comprising the scientific establishment in the new academies.

However, as always, there were exceptions. Jean-Baptiste Du Hamel, the first permanent secretary of the Académie Royale des Sciences, pictured the chemical approach as one of the legitimate roads to scientific truth in his *De consensu veteris et novae philosophiae* of 1663, which went through several editions. And in his popular *La Chymie Naturelle* (1681 and later), Daniel Duncan extended chemical explanations of physiological phenomena beyond those of all previous authors. The practical virtues of chemical medicine continued to be recommended by many authors. Here we have noted James Thompson, who followed van Helmont both in his discussion of fevers and in his attack on bloodletting (1657). And Simeon Partliz (1625; English translation, 1654) attacked the traditional apothecaries and demanded that they employ the new chemical remedies. A final example of practical chemical medicine was found in Christopher Irvine's comprehensive discussion of cure by sympathy (1656).

However, more important than these examples was the development of acid-alkali theory by Otto Tachenius. Here we have already seen the Helmontian background and its development by Sylvius. But Tachenius did not follow the chemical connection with anatomy, as we noted before. In the *Hippocrates Chymicus* (1666), he gave little credit to Paracelsus or van Helmont and ascribed the origins of acid-alkali theory to Hippocrates. Going far beyond the views of others, Tachenius proposed a chemical system of natural philosophy based on acid and alkali as well as a system of medicine.

The *Hippocrates Chymicus* was not a long book, but it was widely read and commented on. In his commentary on the *Pharmoacopoeia Augustana* (1672), Johannes Zwelfer included a long attack on the views of Tachenius, while Robert Boyle rejected the "Hypothesis of Alkali and Acidum" in a section of his work . . . *the Mechanical Origine or Production of divers Qualities* (1675). Boyle, in turn, was taken to task by François de Saint-André in the second edition of his *Entretiens sur l'acide et l'alkali* (1677), in which he strongly supported the new system. Six years later Dr. Bertrand attacked Tachenius directly and denied the presence of acid and alkali in all bodies, while questioning their value in any medical system. But Bertrand's views were unconvincing for François Aignan (1693), who accepted Hippocrates as the founder of the acid-alkali theory without reservation. Here, once again, we meet an author who was convinced that chemistry was the basis of all physiological processes. The essential conflict of acid and alkali resulted in fermentation, which was essential for the furtherance of any organic–or inorganic–growth or change. And in his *Le Prestre Medecin* (1696), he returned to another Paracelsian theme, that the true physician could be compared with a priest.

In England Saint-André's apology for acid-alkali theory was published in translation in 1689 and seven years later John Colbatch published his *Physico-Medical Essay Concerning Alkaly and Acid* (1696). Here he took issue with those who ascribed disease to acid by arguing rather that alkali was the origin of disease. He proclaimed himself a chemist and a follower of van Helmont, but he limited himself to medicine rather than exploring the whole realm of natural philosophy.

The work of Colbatch led to an unexpected number of replies. Richard Boulton, an editor of Boyle's work, could not understand why anyone would return to acid-alkali explanations after they had been shown to be unsatisfactory by Robert Boyle (1698). However, for the most part the opponents of Colbatch faulted him for choosing alkali rather than acid as the cause of disease. Thomas Emes presented a refutation of Colbatch by way of a dialogue between Mr. Acid and Mr. Alkali (1698), while William Coward complained of Colbatch's meager education in his *Alcali Vindicatum.* These authors all discussed the example of scurvy, which had been emphasized by Colbatch, as well as the role of fermentation in the progress of disease. It is clear that this development of the chemical philosophy was centered on medical problems.

The examples chosen here represent only a small part of the medico-chemical literature of the period. A rapid survey of the book reviews in the *Journal des Sçavans* and the *History of the Works of the Learned* in this period give some indication of the interest in chemistry as a basis of natural philosophy by the close of the seventeenth century. At this time the success of the mechanical philosophy in the physical sciences led many physicians to see this as the path to future success in medicine. It is to this debate between iatrochemists and iatrophysicists that we will turn next.

CHAPTER 5

The Medical War Between Chemists and Mechanists

Scarce had the practice of Physick arriv'd at the State of Tranquillity, when the upstart Paracelsus, *backed by all the gang of chymists, raised a new Tumult, and laid fresh cloggs upon the true Practise. Nor was this long past, when Helmontius with his Retainers inlarged its troubles, in so much that all the Conclusions of natural things were chain'd down to a Chymical Furnace.*[1]

Giorgio Baglivi [(1696); English translation, *The Practice of Physick* (1704), p. 123]

There is ample evidence to show the continued interest in chemical interpretations of physiology and chemical pharmaceuticals in the closing decades of the seventeenth century as well as the opening years of the new century. And although there was a sharp debate between "ancients" and "moderns,"[2] there was at least in medicine a declining reliance

[1]Giorgio Baglivi [(1696); English translation, *The Practice of Physick* (1704), p. 123].

[2]Comparisons of ancient and modern learning are common in this period. An entertaining introduction to the debate is to be found in Jonathan Swift's *The Battle of the Books* (1704). The recent studies of Joseph M. Levine are valuable: *Dr. Woodward's Shield: History, Science, and Satire in Augustan England* (Berkeley, Los Angeles, London: University of California Press, 1977); *The Battle of the Books: History and Literature in the Augustan Age* (Ithaca and London: Cornell University Press, 1991). Bernardino Fantini has recently discussed the relation of the chemical-mechanical debate in medicine to the contemporary understanding of living organization. "Iatrofisica e Iatrochimica: La Fisica, La Chimica ed I Modelli dell'Organizzazione del Vivente," *Medicina nei Secoli: Arte e Scienza (Journal of History of Medicine)*, *9* (1997), 59–95.

on ancient authors. Rather than Aristotle and Galen, this new school of physicians turned to mathematics and physics as a proper basis for a reformed medicine. The mathematical approach in astronomy had led from Copernicus to Tycho and Kepler, to Galileo, and finally to Isaac Newton, whose monumental *Principia Mathematica* (1687) presented the physical sciences with a new physics of motion based upon a sun-centered cosmology.

MECHANISM IN MEDICINE IN THE LATE SEVENTEENTH CENTURY

The move toward a mechanistic view of nature and man began to affect both chemistry and medicine in the second half of the seventeenth century. We see this in a number of authors. In the course of two years Walter Charleton (1620–1701) moved from an Helmontian view of nature in which he repeatedly cited authors such as Digby, Severinus, Hartmann, Kircher, Porta, and Fludd[3] to a mechanistic one, now referring his readers to Copernicus, Gilbert, Galileo, Mersenne, Descartes, and Harvey.[4] By 1654 Charleton openly attacked Paracelsus, Crollius, and van Helmont in his *Physiologia Epicuro-Gassendo-Charletoniana*.[5] And we have already seen evidence of Robert Boyle's early reliance on the works of van Helmont supplanted (c. 1658) by his later realization that he had once been overly influenced by those who rejected a mathematical basis for the sciences in favor of an emphasis on the study of matter.[6]

William Harvey's quantification of the blood flow in the *De motu cordis* (1628) is frequently offered as an early example of the quantification of the biological sciences, as is the data gathered by Sanctorius (1561–1636) in his

[3]Nina Rattner Gelbart, "The Intellectual Development of Walter Charleton," *Ambix*, *18* (1971), 149–168; Allen G. Debus, *The Chemical Philosophy: Paracelsian Science and Medicine in the Sixteenth and Seventeenth Centuries*, 2 vols. (New York: Science History Publications, 1977), *2*, pp. 469–473; J. B. van Helmont, *Ternary of Paradoxes. The Magnetick Cure of Wounds. Nativity of Tartar in Wine. Image of God in Man*, translated, illustrated and ampliated by Walter Charleton (London: James Flesher for William Lee, 1650), sig. 3 verso.

[4]Gelbart, "Intellectual Development," 160.

[5]Walter Charleton, *Physiologia Epicuro-Gassendo-Charltoniana* . . . , Introduction and indexes by Robert Hugh Kargon (New York/London: Johnson Reprint, 1966), p. 58.

[6]See above, chapter 3. Robert Boyle, *The Usefulness of Experimental Philosophy*, Part 2, Section 2, in *The Works of the Honourable Robert Boyle*, 6 vols. (London: J. and F. Rivington *et al.*, 1772), vol. 3, p. 426.

Ars de statica medicina (1614).[7] Here he weighed the body under many conditions, before and after meals, while sleeping and awake, etc. But there was also an interest in mechanical analogies, as when Harvey compared the heart to a bellows[8] or when Boyle spoke of the body as an "engine" and suggested that some diseases such as seasickness resulted from motion, a mechanical cause.[9] And we have already seen that George Castle compared the human body to clockwork in his *The Chemical Galenist* (1667).[10] For him this was far more satisfactory than the explanations of the iatrochemists.

Perhaps more important for the development of a mechanical physiology on the Continent was the work of René Descartes (1596–1650), Giovanni Alfonso Borelli (1608–1679), and Lorenzo Bellini (1643–1704). Although Descartes is praised for his philosophy and its effect on the physical sciences in the late seventeenth and early eighteenth centuries, he was well aware of the occult interests of his age. His youthful interest in the Rosicrucian tracts has been the subject of recent scholarly interest,[11] and in his *Discourse on Method* (1637) he recalled that as a student he "had gone through all the books I could lay my hands on dealing with the occult and rare sciences."[12]

And if today we emphasize Descartes' stress on mathematical logic as a key to understanding nature, he openly argued that the true goal of science is "principally for the preservation of health, which is undoubtedly the first good, and the foundation of all other goods of this life; . . . if it is possible to find some other means of rendering men as a whole wiser and more dexterous than they have been hitherto I believe it must be sought in medicine."[13] For this reason he resolved "to devote the time I have left to live to no other occupation than that of trying to acquire some knowledge of Nature, which

[7]A. Wolf, *A History of Science, Technology, and Philosophy in the 16th & 17th Centuries*, New Edition prepared by Douglas McKie (London: George Allen & Unwin Ltd., 1950), pp. 431–435.

[8]Walter Pagel, *William Harvey's Biological Ideas: Selected Aspects and Historical Background* (Basel/New York: S. Karger, 1967), pp. 212–214.

[9]Robert Boyle, "Some Considerations Touching the Usefulness of Experimental Natural Philosophy," Part 2, Section l, in *Works*, vol. 2, p. 180.

[10]George Castle, *The Chemical Galenist: A Treatise Wherein The Practice of the Ancients is reconcil'd to the new Discoveries in the Theory of Physick* . . . (London: Sarah Griffin for Henry Tinyford, 1667), pp. 5–6.

[11]William R. Shea, "Descartes and the Rosicrucians," *Annali dell'Istituto e Museo di Storia della Scienza di Firenze*, *4* (1979), 29–47.

[12]René Descartes, *Discourse on Method and Other Writings*, Translated with an Introduction by F. E. Sutcliffe (Baltimore: Penguin Books, 1968), p. 29.

[13]*Ibid.*, pp. 78–89.

may be such as to enable us to deduce from it rules in medicine which are more assured than those we have had up to now; . . ."[14]

Although Descartes was never a practicing physician, there is little doubt that his work indicates a considerable interest in anatomy and physiology. In the *Discourse on Method*, he described the motion of the heart and blood

> which follows as necessarily from the mere disposition of the organs which may be observed in the heart by the eye alone, and from the heat which one can feel there with one's fingers, and from the nature of the blood which can be known by experience, as does that of a clock from the power, the position and the shape of its counter-weights and its wheels.[15]

He thought of the continual motion of the blood in terms of a mechanical analogy and he referred his reader to the work of "an English doctor" [Harvey] who had shown that the blood circulated perpetually in the body.[16]

For Descartes the complexity of the body, its motions, senses, and even its internal passions are best explained mechanically.

> This will not appear in any way strange to those who, knowing how many different automata or moving machines the industry of man can devise, using only a very few pieces, by comparison with the great multitude of bones, muscles, nerves, arteries, veins and all the other parts which are in the body of every animal, will consider this body as a machine, which, having been made by the hands of God, is incomparably better ordered, and has in it more admirable movements than any of those which can be invented by men.[17] . . . [And indeed] one may very well liken the nerves of the animal machine I have described to the pipes of the machines of those fountains; its muscles and its tendons to the other different engines and springs that serve to move them; and its animal spirits, of which the heart is

[14]*Ibid.*, p. 91. Medical historians have generally emphasized the mind-body problem in their discussions of Descartes. See, for instance, Richard B. Carter, *Descartes' Medical Philosophy: The Organic Solution to the Mind-Body Problem* (Baltimore and London: The Johns Hopkins University Press, 1983).

[15]Descartes, *Discourse on Method*, p. 68.

[16]*Ibid.*, p. 69.

[17]*Ibid.*, p. 73.

> the source and the ventricles of the brain the reservoirs, to the water that moves these engines. Moreover, respiration and other similar functions which are usual and natural in the animal machine and which depend on the flow of the spirits are like the movements of a clock or of a mill, which the ordinary flow of water can make continuous.[18]

More important for the development of a mechanical physiology was the *De motu animalium* of Giovanni Alfonso Borelli (1608–1679).[19] Influenced by Galileo, Borelli sought to apply the principles of mathematical physics to the problem of human motion. He described muscular motion in terms of levers and weights and also discussed lifting, running, jumping, and a wide variety of human and animal movements. Although Borelli's mathematical treatment is dominant, it is evident that he was also aware of current chemical thought. Thus, he discussed the separating action of fermentation.[20] And, referring to van Helmont's willow tree experiment, he concluded that "it does not seem unlikely that plants are fed by and grow from water alone as if it was blood."[21]

In the case of Lorenzo Bellini (1643–1704), we find a brilliant youth recognized early in life.[22] At the University of Pisa he studied under Borelli and seemed so accomplished that he was appointed professor of theoretical medicine there at the age of twenty. Five years later he accepted the chair of

---

[18]Descartes, *Oeuvres*, edited by Ch. Adam and P. Tannery, vol. 11, pp. 131–132 as cited by A. Rupert Hall in *From Galileo to Newton 1630–1720* (New York and Evanston: Harper & Row, 1963), p. 193. Hall discusses Descartes' mechanistic views in *ibid.*, pp. 192–197. A major text for understanding mechanism in the biology of this period is Mirko D. Grmek, *La première révolution biologique: Reflexions sur la physiologie et la medecine du xviie siècle* (Paris: Editions Payot, 1990). See also Theodore M. Brown, "Physiology and the Mechanical Philosophy in Mid-Seventeenth Century England," *Bulletin of the History of Medicine, 51* (1977), 27–54. A study of special interest is Emilio Balaguer Perigüell, *La Introducción del Modelo Físico-Matemático en la Medicina Moderna: Análisis de la obra de G.A. Borelli (1608–1679) De motu animalium, Cuadernos Hispánicos de Historia de la Medicina y de la Ciencia, XIV*, Serie A (Monografías) (Valencia-Granada, 1979).

[19]The *De Motu Animalium* is conveniently available as Giovanni Alfonso Borelli, *On the Movement of Animals*, translated by Paul Maquet (Berlin, Heidelberg et al.: Springer-Verlag, 1989). See also the works of Hall and Grmek listed above in note 18.

[20]Borelli, *De Motu*, pp. 341–342.

[21]*Ibid.*, p. 387.

[22]On Bellini see the article by Theodore M. Brown in the *Dictionary of Scientific Biography*, Editor in Chief, Charles Coulston Gillispie, 16 vols. (New York: Scribner's, 1970–1980), vol. 1 (1970), pp. 592–594. Bellini's views on respiration are discussed by J. R. Partington, *A History of Chemistry*, 4 vols. (London: Macmillan, 1962–1970), vol. 2, pp. 450–451.

GIOVANNI ALFONSO BORELLI

dal dipinto ad olio esistente sul suo Monumento sepolcrale in Roma.

Portrait of Giovanni Alfonso Borelli (1608–1679). Courtesy of the National Library of Medicine.

anatomy and late in life he became the first physician to Duke Cosimo III of Tuscany.

Bellini attracted attention with the publication of a study of the kidneys, his *Exercitatio anatomica de usu renum* (1662), in which he rejected Galenic explanations and explained the formation of the urine with the aid of mechanical views. Instead of a Galenic "faculty" in the kidneys, Bellini pointed to the structure of the parenchyma with its "fibers, open spaces and densely packed tubules opening into the pelvis of the kidney."

> Bellini posited that arterial blood brought into the kidneys is separated into urinous and sanguineous portions by a sievelike arrangement in the renal vessels and fibers: the urinous portion collected in the urinary tubules and the blood returned to the circulatory stream through the venules. He meant this mechanical explanation of urinary secretion, achieved by the "configuration alone" of the renal vessels, as a general substitute for Galen's incomprehensible "faculties."[23]

Bellini returned to the study of urine in 1683. Here, although influenced by Willis, he rejected the English physician's reliance on fermentation, preferring a mechanical explanation of the blood. He was to expand his views in a final set of essays, the *Opuscula aliquot*, in 1695.

By the closing years of the century the move toward iatrophysics, that is, a mechanical medicine, was in full swing. The prefatory essay to the first volume of the *Histoire* and *Memoires* of the French Academy (1699) praised mathematics and the physical sciences. "The geometric spirit is not so exclusively attached to Geometry that it could not be drawn upon or transported to other areas of knowledge. A work on morals, on politics, on criticism, perhaps even eloquence, will be better, all other things being equal, if it is prepared with the hand of Geometry."[24] The biological sciences are not so advanced, but anatomy may equal astronomy for the person who seeks truth of his Creator. "Above all, Astronomy and Anatomy are the two sciences which offer us most visibly two grand characteristics of the Creator; the first, his Immensity which may be seen by the distances, the grandeur and the number of celestial bodies, the other, his infinite Intelligence, by the mechanics of

---

[23]Brown, "Bellini" in *DSB*, *1*, p. 453.

[24]L'Academie Royale des Sciences (Paris), *Histoire* (1699), (2nd edition, Amsterdam: Pierre Mortier, 1734), "Preface," pp. xvii–xviii.

animals. True natural philosophy rises to the level of becoming a type of Theology."[25] Chemistry is surely less admired in this essay. To be sure, the skilled chemist may prepare mineral remedies valuable for the physician, but no one should place faith in the mystical concepts of traditional alchemy.[26] Little reference was made here to the chemical medicine of the past half century.

BAGLIVI AND PITCAIRNE

The search for a new mechanical medicine becomes ever more evident as we turn to other prominent physicians of the period. One of these, Giorgio Baglivi (1668–1707)[27] took his M.D. degree at Salerno (1688) and studied also at Naples and Pisa, where he attended the lectures of Bellini before working in hospitals throughout Italy. He settled in Bologna in 1691 as assistant to Marcello Malpighi, whom he followed to Rome the next year. After the death of Malpighi in 1694, Baglivi was recognized as a leading physician on his own. He became the Pope's second physician and was elected professor of anatomy and then professor of theoretical medicine at the Sapienza in Rome. In 1696 Baglivi published his *De praxi medica*, which went through many Latin editions and was translated into English as *The Practice of Physick* in 1704. Here we see his distrust of medical systems and his emphasis on reason and observation: "the two chief Pillars of Physick are *Reason* and *Observation*: But Observation is the Thread to which Reason must point."[28] Baglivi claimed to oppose all systems in medicine. Commenting on the general rejection of the ancient authors, he observed that "Those who gave the first vent to this Error, I take to be *Helmont* and *Paracelsus*, who in order to raise their own Sect upon the ruines of the *Galenick*, not only charg'd *Galen* and almost all the Ancients with Error and Ignorance, but lash'd 'em with Calumnies and Taunts, in their Medicinal Writings."[29] Indeed,

---

[25] *Ibid.*, p. xxi.

[26] *Ibid.*, p. xii.

[27] See Mirko D. Grmek, "La vita e lopera di Giorgio Baglivi, medico raguseo e leccese," in *Il nucleo filosofico della scienza*, Guido Cimino et al., eds. (Galatina: Congedo, 1991), pp. 93–111 and the biographical account by Mirko Grmek in the *Dictionary of Scientific Biography*, vol. 1 (New York: Scribner's, 1970), pp. 391–392. An essential collection of twelve articles on Baglivi appeared after the completion of this work in *Medicina nei Secoli: Giornale di Storia della Medicina, 12* (2000).

[28] Giorgio Baglivi, *The Practice of Physick, Reduc'd to the Ancient Way of Observations Containing a Just Parallel between The Wisdom and Experience of the Ancients, and The Hypothesis's of Modern Physicians* . . . , trans. George Sewell and J. T. Desaguliers (2nd ed., London: Andr. Bell et al., 1704), p. 9.

[29] *Ibid.*, p. 20.

Portrait of Giorgio Baglivi (1668–1707). Frontispiece to his *Opera omnia Medico-Practica et Anatomica* (10th edition, Antwerp, 1734). From the collection of the author.

> Scarce had the Practice of Physick arriv'd at the State of Tranquillity, when the upstart *Paracelsus*, backed by all the gang of Chymists, raised a new Tumult, and laid fresh cloggs upon the true Practise. Nor was this long past, when *Helmontius* with his Retainers inlarged its troubles, in so much that all the Conclusions of natural things were chain'd down to a Chymical Furnace.[30]

It is true that medicine has been enriched by chemically prepared remedies, but "those who reason from the effects of Minerals in a chymical Furnace, to the effects observed in the Body of an Animal, do not only conclude in a lame imperfect way, but give occasion in a Flaming degree to those pernicious Errors, which unjustly disturb our Modern Physick . . ."[31] But the Chemical Philosophy was not the only problem. A whole host of philosophies have arisen: Cartesian, Democritean, Mechanical, Physico-Mechanical, and others as well. The result has been that physicians have become philosophers "and being allur'd with the tranquillity of Attractive Wisdom, raised up a contentious and wrangling sort of Philosophy in the Room of the Practice of Physick . . ."[32]

Although opposed to the philosophies adopted by physicians, Baglivi found himself attracted by the mechanists and he admitted that practical physicians should follow the example of astronomers in forming the hypotheses of diseases. And he noted that

> Since Physicians began to examine the Structure and Actions of a living Body, not by Physico-Mechanical and Chymical Experiments, but by Geometrico-Mechanical Principles, they have not only discover'd an infinite number of things that were unknown to former Ages; but have made it out, That a Human Body, as to its natural Actions is truly nothing else but a complex of Chymico-Mechanical Motions, depending on such Principles as are purely Mathematical. For whoever takes an attentive view of its Fabrick, he'll really meet with Shears in the Jaw-bones and Teeth, a Phiol in the Ventricle, Hydraulick Tubes in the Veins, Arteries, and other Vessels, a Wedge in the Heart, a Sieve or Straining-holes in the Viscera, a Pair of Bellows in the Lungs, the Power of a Leaver in the Muscles, Pulleys in

---

[30] *Ibid.*, p. 123.

[31] *Ibid.*, p. 36. See also the *History of the Works of the Learned*, *1* (1699), 260.

[32] Baglivi, *Practice*, p. 125.

Portrait of Archibald Pitcairne (or Pitcairn) (1652–1713). Courtesy of the National Library of Medicine.

> the Corners of the Eyes, and so on. And tho' the Chymists explain the *Phaenomena* of natural Things, by the Terms of *Fusion, Sublimation, Precipitation,* &c. And so make a separate sort of Philosophy; yet all these ought to be imputed to the Force of a Wedge, Balance, Leaver, Spring, and such like Mechanical Principles. . . . the natural Effects of an animated Body can't be accounted for with greater Facility and Clearness any other way, than by those Mathematico-Experimental Principles, by which Nature speaks her own Mind; . . .[33]

Hardly less important were the works of Archibald Pitcairne (1652–1713), a Scottish physician whose writings have been the subject of a number of recent papers.[34] A student first in Edinburgh, Pitcairne took his M.D. at the University at Rheims. Back in Edinburgh he became a founding fellow of the Royal College of Physicians of Edinburgh (1681) and then was appointed professor of medicine at the University there (1685). Pitcairne was influenced by the work of Borelli as well as the *Principia* of Newton, and shortly after accepting a position of professor of medicine at the University of Leiden in 1691 he became aware of Huyghen's work on probability. As a mechanist he found Bellini's work of great interest and praised him as early as 1692.[35] For Bellini this open appreciation was so welcome that he dedicated his *Opuscula aliquot* (1695) to Pitcairne. In a work on the circulation of the blood published in 1693, Pitcairne took up the problem of the delivery of particles in the body through the blood to specific locations where they were required. He concluded that through a particle/sieve system some particles would pass while others would not.[36] This type of reasoning was to be expanded in his work on fevers two years later.

---

[33] *Ibid.*, pp. 135–136.

[34] On Pitcairne, see the recent papers by Anita Guerrini, "James Keill, George Cheyne, and Newtonian Physiology, 1690–1740," *Journal of the History of Biology, 18* (1985), 247–266; "The Tory Newtonians: Gregory, Pitcairne, and Their Circle," *Journal of British Studies, 25* (1986); "Archibald Pitcairne and Newtonian Medicine," *Medical History, 31* (1987), 70–83. Pitcairne's relation to Edward Eizat is detailed in Stephen M. Stigler, "Apollo Mathematicus: A Story of Resistance to Quantification in the Seventeenth Century," *Proceedings of the American Philosophical Society, 136* (1992), 93–126. The name is spelled both Pitcairne and Pitcairn. Here we will use Pitcairne except in the original citation of his *Works* where Pitcairn appears on the title page.

[35] See the article on Pitcairne by Theodore M. Brown in the *Dictionary of Scientific Biography*, vol. 11 (New York: Scribner's, 1975), pp. 1–3 (2).

[36] Stigler, 105–109.

Throughout the work of Pitcairne we see a rejection of chemical explanations and a reliance rather on mathematics and mechanics as a true basis for medicine.

> ... Physicians ought to propose the method of Astronomers as a Pattern for their Imitation. These never take up, and adopt into their Science such Opinions as are grateful to the Vulgar, or generally received by Orators: Never in the Explication of the Motion of the Planets, call in the Assistance of a Romantic Hypothesis concerning the Structure of the World, however pleasing and plausible, but by comparing the Observations which have been made at great Distances of Times and Places, and put together in a Method familiar to them, and useful to all the Phaenomena of the Celestial Motions, and so compute the Powers and Force which Bodies in Motion observe in their Tendency to other Bodies, either moveable or immoveable.[37]

We need not dispute questions related to substantial forms, subtile matter, the accidental concourse of atoms, sympathy and antipathy, or even fermentation, since nothing like this will be found in the works of astronomers.[38] Above all, as physicians we should congratulate our age, especially for the work of Isaac Newton "since we may justly hope, that by the assistance of the Principles demonstrated by that Great Man, the Powers and Properties of Bodies serviceable to Medicinal Uses and the Comfort of Mankind, may be discovered with greater Ease, and reduced to a greater Certainty."[39]

How different Newton's work is compared to that of the chemists who–as an example–have sought to explain digestion abnormally and "have introduced into the Stomach, either certain powerful *Daemons*, and exalted invisible *Spirits*, or *Ferments*, and other Fluids, of various Denominations, which are supposed able to dissolve any manner of Substances . . . ,"[40] but all that is needed is a means to reduce food to a fluid and this may be accomplished "without the Assistance of a *Daemon, or a Stygian* Liquor." Neither of these nor "a Fluid abounding with an Acid or Volatile Body, nor a Salt or Acrid, nor a Compound of these or other Particles of any Nature are the

[37]Archibald Pitcairn, *The Whole Works . . . Wherein are Discovered, The True Foundation and Principles of the Art of Physic. With cases and Observations upon most Distempers and Medicines*, trans. George Sewell and J. T. Desaguliers (2nd edition, London: J. Pemberton for W. and J. Innys, 1727), p. 11.

[38]*Ibid.*, pp. 12–17.

[39]*Ibid.*, p. 56.

[40]*Ibid.*, p. 109.

Instruments of the Dissolution and Digestion of Food in the Stomacks of Animals: Much less can this Operation be successfully performed by the invisible Spirits of *Helmont* and *Wedelius*, or the Daemon of Doleus."[41]

But how can any ferment dissolve flesh and not dissolve the fibers of the stomach? Indeed, it is the fibers that prepare the food for nourishment through a preparatory crushing action. ". . . it is only the Motion of the stomach working and comminuting the Food, which finished Digestion by a separation of the last-formed Parts into Pipes and Fibres of such a Nature as is observable in Animals."[42] In short, "there is nothing transacted in the Stomach, from whence one may certainly conclude that Salts, Sulphurs, or other Bodies, that pass for Principles, can be extracted from the Concoction of the Food . . ."[43] There is no Nutritive Sulphur made in the Stomach and chylification is not due to an invisible spirit . . . nor is an analysis of alkali and acid responsible for relaxing the union of the sulphur in the stomach. All of this chemical nonsense regarding digestion–and here Pitcairne includes a sharp attack on the doctrine of acid and alkali[44]–is to be discarded because "the Food cannot be concocted without the grinding muscles of the *Abdomen*."[45] In short, digestion–and excretion as well–was best explained for Pitcairne in terms of the expansion and contraction of the fibers of the body.

EARLY EIGHTEENTH-CENTURY IATROCHEMISTRY

In spite of the prominence of Baglivi, Bellini, and Pitcairne, it would be wrong to assume that there were no defenders of chemistry. The eighteenth century was to witness the continued publication of collected alchemical works such as the vast collections in Latin of Manget, in German of Roth-Scholz, and in French of Richebourg. But in addition, there remained a strong interest in iatrochemistry. For example, J. Mongin produced his *Le chimiste physicien* in 1704. Here he lauded the modern discoveries that had overturned the ancients.[46] Above all, he turned to the work of van Helmont and Descartes, to whom he gave joint credit for the new science. It was, however, van Helmont

[41] *Ibid.*, p. 110.

[42] *Ibid.*, p. 117.

[43] *Ibid.*, p. 134.

[44] See the *History of the Works of the Learned, 4* (1702), 150.

[45] Pitcairne, p. 136.

[46] Mons. J. Mongin, *Le Chimiste Physicien* . . . (Paris: Laurent d'Houry, 1704), sig. ãv verso.

to whom he gave most credit because he felt that the future of medicine must rest upon a chemical examination of the body and the "nature and uses of the liquors which revolve there without ceasing."[47] This research would surely lead to a determination of how the functions of the body are maintained, and it would follow other discoveries based on chemistry that have recently given us a clear and distinct idea of the dissolution of foods, of the change in the chyle and the blood, of the nourishment and the growth of the bodily parts, and of the various filtrations that occur in the body.

Even in the French Academy of Sciences, that stronghold of mathematical investigation, there were defenders of chemistry. Martino Poli (1660–1714), an Italian who had been elected an extra foreign associate of the Academy in 1702, published his *Triumph of Acid vindicated from the calumnies of many moderns (Il Trionfo degli Acidi vendicati dalle calunnie di molti Moderni)* (1706). Here Poli took aim at those physicians who adhered to the use of mechanical principles. Rather, he sought openly to demonstrate the absurdity of the mechanical philosophy. Here his targets were Descartes, Gassendi, and their disciples.

He began with a discussion of fermentation and the elements. Like Paracelsus and many others, he did not reject the four elements of the Aristotelians. However, he did criticize the acid-alkali system of Otto Tachenius and the five principles of Thomas Willis. Poli preferred the three principles of the Paracelsians who were to him the true Hermetic chemists. In contrast to other systems, salt, sulphur and mercury had been "confirmed by experience."[48]

Poli discussed unfavorably the Cartesian view of the universe and then turned to the case of living bodies. These, he felt, could not be discussed in terms of inert ones. And since this was true, it was difficult to understand why the mathematicians and the mechanists continued their application of gross mechanics and analogies borrowed from machines. The attempt to explain digestion in terms of fibers and motion was absurd.[49] These authors might speak of filters transmitting some liquors and excluding others (as did Pitcairne), but the passage of the chyle in the lacteal veins is not the result of a mechanical filtration or of any purely mechanical cause, nor are the separations of urine in the reins or the bile in the liver to be explained mechanically. Rather, these operations are due to chemical processes: fermentation,

---

[47] *Ibid.*

[48] Martino Poli, *Il Trionfo degli Acidi vendicati dalle calumnie di molti Moderni* . . . (Rome: Giorgio Placho, 1706) as reviewed in the *Journal des Sçavans, 38* (1707), p. 333.

[49] Poli, *Il Trionfo*, pp. 142–144.

dissolution, sublimation, "and other similar operations which occur in the body of an animal as in the laboratory of a chemist."[50]

THE DEBATE OVER DIGESTION

A sharp debate occurred in the medical literature between 1709 and 1715. This was chronicled in Jean Astruc's (1684–1766) *Traité de la cause de la Digestion* (1714). In his account, Astruc referred to mechanical descriptions of digestion given by Leeuwenhoek in 1684 and by Pitcairne in 1693. However, Astruc continued, neither of these accounts greatly affected the current debate that had begun with an article by the mechanist, Philippe Hecquet (1661–1737), in 1709. Both Astruc and Raymond Vieussens replied to Hecquet the following year and in 1712 Hecquet published his *Traité de Digestion*, which was answered by Astruc in 1714.

Here we need only sample the works of these authors to obtain some sense of the issues being discussed. Philippe Hecquet graduated in medicine at Rheims (1684) and then took his M.D. at Paris ten years later. A convinced mechanist, he presented his views in a *Traité des Dispenses du Carême* (1709), which he followed with *De Digestion et des Maladies de l'Estomac; Suivant le systeme de la* Trituration *& du* Broyement, *sans l'aide des Levains ou de* Fermentation, *dont on fait voir l'impossibilité en* santé *& en* maladie (1712). Here and in later works [*De Digestion . . .* (Nouvelle edition, revised, corrected and augmented, 1730); *Le Brigandage de la Chirurgie* and *Le Brigandage de la Medecine Pharmacie* (1738); *Le Brigandage de la Medecine* (1733, reprinted 1749)], we can follow his views on chemistry and mechanism in medicine.

For Hecquet the chemical physicians had effectively declared war on the true medicine.[51] They had neglected the study of anatomy and quite incorrectly substituted for it analysis by fire. But chemistry is more a manual art

---

[50] *Journal des Sçavans, 38*, 336–338.

[51] Philippe Hecquet, *Le Brigandage de la Medecine, dans la maniere de traiter les petites véroles & les plus grandes maladies par l'Emétique, la Saignée, du pié & le Kermès minéral . . .* (3 vols., Utrecht: Corneille Guillaume le Febure, 1749; 1st edition, 1733), vol. 1, p. 10. L. W. B. Brockliss has studied the relation of Hecquet's iatrophysical views with his Jansenist beliefs. He was a convinced mechanist who rejected the views of several iatrochemists in print. As he grew older he became ever more suspect of chemically prepared medicines, but he remained convinced that God expected man to seek truth in the book of nature and that medical cure occurred only in cooperation with the Holy Spirit. Brockliss, "The Medico-Religious Universe of an Early 18th Century Parisian Doctor: The Case of Philippe Hecquet" in *The Medical Revolution of the 17th Century*, edited by Roger French and Andrew Wear (Cambridge: Cambridge University Press, 1989), pp. 191–221.

Portrait of Philippe Hecquet (1661–1737). Courtesy of the National Library of Medicine.

than a true science. Fire only decomposes substances through distillations, fusions, reverberations, calcinations, and other processes that destroy the tissue or the natural arrangement of the parts that compose a body.

It is clear that a chemist will never make a good physician.[52] Rather, geometry and mechanics are the basic laws required by a physician . . . precisely those laws which are the basis of natural philosophy and have no need of chemical fires. It is true that this subject has presented medicine with some useful preparations, but chemistry is an "art more mechanical or manual than scientific . . . more fit for artisans than philosophers."[53] Indeed, traditional Galenicals are to be preferred to chemically prepared medicinals. Because of the ill effects of fire, Hecquet recommended to those who would study chemistry a recent book by Ernest Gotthold Struve that avoided fire and heat as agents and sought the preparation of substances by other means: mixing, light, natural motion, solution, deliquation, crystallization, digestion, and precipitation.[54] In this book Struve concluded that pharmaceutical chemistry was equally productive with or without fire.[55]

But what are physicians to turn to instead of chemistry? Here Hecquet wrote specifically against Vieussens. No less than Pitcairne, Hecquet rejected the reliance of physicians on systems–and of all systems none seemed worse to him than that of fermentation and the views of the chemists.[56] The true answer was to be found in trituration, which avoids many inconveniences of the views of the chemists. It is easy and simple and like the thread of Ariadne leads to truth. Instead of heat, fire, and effervescence, we deal with the oscillations of fibrous matter. The chemists talk of dryness, concentrations, coagulations, and the thickening of liquors, while we speak of pressures. With them we have many salts, tartarous and sticky fluids, while here we have the effect of diameters and sluggish oscillations. Chemists write of bizarre fermentations, while we discuss disordered fibers. It follows that in these chem-

---

[52]Hecquet, *Brigandage de la Medecine, 1*, p. 124.

[53]Philippe Hecquet, *Le Brigandage de la Chirurgie . . . Premiere Partie* [and] *Le Brigandage de la Pharmacie . . . Seconde Partie* (Utrecht: Chez les Soeurs de Corneille-Guillaume le Fevre, 1738), *Seconde Partie*, p. 63.

[54]Ernest Gotthold Struve, *Paradoxum Chymicvm sive Igne i.e. Operationes et Experimenta Physico-Chymico-Pharmacevtica, ipsaqve Medicamenta Chymica Ignis Operari Solita Sive Igne* (Jena: Ernest. Clavd. Bailliae, c. 1715), pp. 55–61 and *passim.*

[55]*Ibid.*, p. 80.

[56]Philippe Hecquet, *De la Digestion et des Maladies de l'Estomac, suivant le Systeme de la Trituration & du Broyement, sans l'aide de levains ou de la fermentation, dont on fait voir l'impossibilité en santé & en maladie* (Paris: François Fournier, 1712), pp. x–xi.

ical systems of the bile, the saliva, the blood, and melancholy, of acid and alkali, volatility, fixity, aqueous, sulphureous, spiritous and phlegmatic, that all that is really needed is a knowledge of solid and liquid, and in place of all the chemical faculties, qualities and flavors, we really only need resistances and forces. And finally, he added, because one does not want to repeat all these names too often, we insist that all these qualifications are imaginary, while solids and liquids are natural and in nature.[57]

Too many otherwise honest physicians have wasted their time and efforts on chemistry. Surely this was true of Willis and Sylvius, both of whom devoted themselves too much to the dreams of Paracelsus and van Helmont. There are too many French physicians such as Daniel Duncan who even now present their medical studies in terms of chemistry.[58] Others idolize the works of Paracelsus, Libavius, van Helmont, and the whole tribe of alchemical adepts. Unless this fascination is eliminated, medicine will degenerate to a monstrous science.

Although Hecquet was a friend and supporter of Pitcairne, he was strongly opposed to the views of Raymond Vieussens (ca. 1635–1715), one of the most distinguished French physicians of his time. After taking his M.D. at Montpellier in 1670, he became a physician at the main hospital there and rose to the rank of chief physician. He is renowned for his anatomical studies and he relates that it was his reading of Willis in 1671 that led to his study of the nerves and the publication of his *Neurographia universalis* in 1684. He then turned to an investigation of mixed bodies and the chemistry of the human body, a work to which he added a treatise on fermentation [*Nouveau Système des Liqueurs du Corps humain Vaisseaux*, 1705].[59] Of particular interest for the development of iatrochemistry is Vieussens' discussion of fermenta-

[57]*Ibid.*, pp. xxix–xxx.

[58]Philippe Hecquet, *De la Digestion et des Maladies de l'Estomac; Suivant le systeme de la Trituration & du Broyement sans l'aide des Levains ou de la Fermentation, dont on fait voir l'impossibilité en santé & en maladie* (Nouvelle ed., revised, corrected and augmented by the author, 2 vols., Paris: Guillaume Cavalier, 1730), vol. 2, p. 7.

[59]On Vieussens, see the article with references by M.D. Grmek in the *Dictionary of Scientific Biography*, *14* (1976), pp. 25–27. Vieussens wrote of his life and scientific progress in the preface to his *Deux dissertations . . . Le Premiere Touchant l'extraction du Sel acide du Sang, La Second Sur la proportion de quantité de ses principes sensibles* (Montpellier: Honoré Pech, 1698). For his early work on the nervous system and on fermentation he was awarded an annual pension of 1,000 livres by the king and later named as a councillor of state. See above sig. ãi^{r}. He spoke further of his research in the *Traité nouveau des Liqueurs du Corps Humain* (1705). See the *Traité nouveau de la Structure et des Causes du mouvement natural du Coeur* (2 vols., Toulouse: Jean Guillemette, 1715), vol. l, pp. ê iv$^{r\&v}$; ii^{r}.

Portrait of Raymond Vieussens (ca. 1635–1715). Courtesy of the National Library of Medicine.

tion in his *Traité de la structure du Coeur.*[60] Here he assures the reader that blood contains both passive and active principles of natural fermentation and that the fermentation process takes place in the ventricles of the heart. The first principle is composed of a volatile sulphur and an acid salt. The second is the active principle of natural fermentation of the blood and nothing else but a spiritous substance impregnated with volatile particles of an acidic salt. "I wish to say that the vital animal spirit is united to the animal spirit . . . and by vital spirit, I understand a very fine liquor, present throughout all of the blood and principally composed of very subtle air, charged with volatile nitrous particles . . ." He goes on to describe the course of this vital spirit through the nerves, veins and arteries, indicating that it finally reunites with the animal spirit sustaining the fermentation process. It takes on its red color from the nitrous particles united to volatile acid salts received from food. Vieussens considered his study of blood his most important work, and indeed, his identification of the vital spirit is the continuation of a century of Paracelsian and chemical investigations seeking the isolation of this substance.[61]

Although not normally pictured as an important figure in the history of chemistry, it is clear that Vieussens was influenced strongly by the post-Helmontian authors, Willis and Sylvius. However, he was also indebted to Descartes for his views on the elements.[62] In the *Traité nouveau des Liqueurs de Corps Humain*, he described the three elements of Descartes–the first, small and moving with great velocity; the second, heavier and slower; and the third, the heaviest, of irregular form and very slow. These he equated with earlier thought, the first with the fire of the ancients, the Platonic soul of the world, and the Hippocratic igneous matter. The second element was an aetherial matter and the third was simply earth.

If then we consider Vieussens as a late seventeenth-century chemical physician as well as an anatomist, we will find it easier to understand his position in the debate over digestion. In January, 1710 he published "De la nature & des proprietez du Levain de l'Estomac in *Des Mémoires de Trevoux.*[63] This was a reply to the iatrophysicists, especially to Hecquet's *Traité des Dispenses*

[60] *Ibid.*, *1*, pp. 120–124.

[61] Allen G. Debus, "The Paracelsian Aerial Niter," *Isis*, *55* (1964), 43–61; Allen G. Debus, "Chemistry and the Quest for a Material Spirit of Life in the Seventeenth Century," *Spiritus: IV Colloquio Internazionale del Lessico Intellettuale Europeo, Roma, 7–9 gennaio 1983*, M. Fattori and M. Bianchi, editors (Roma: Edizioni dell'Ateneo, 1984), pp. 245–263.

[62] Vieussens, *Traité nouveau de la Structure*, vol. 2, p. 2.

[63] Hecquet (1730), vol. 2, pp. 507–520.

du Carême, published the previous year. In this article Vieussens refuted those who argued that the coction of foods begins mechanically in the mouth by biting and chewing and then continues in the stomach, which is viewed as a hollow muscle that kneads the food particles and finally dissolves them by the force of moving fibers. In this mechanical fashion the food is turned into a fine and delicate cream.

Vieussens replied that such an explanation was not accepted by physicians who understand the perfect natural structure and function of the stomach. As for the mouth, the action of the teeth is quite different from that of the stomach and the chemical action of the saliva is a significant factor in the digestive process. It is true that there is a motion of the stomach, but it is not a crushing and grinding action as Hecquet would have you believe. Rather, it is gentle and cannot explain the reduction of the food into a cream-like consistency.[64] One can only conclude that digestion is accomplished by a fermentation process and that the ferment is composed of volatile parts, of phlegm, of sulphur, and of an acid salt.[65] In short, digestion is basically chemical, not mechanical. Needless to say, Hecquet remained a convinced mechanist and when he expanded his 1712 work on digestion to two volumes in 1730 he reprinted Vieussens' 1710 article in its entirety to emphasize the truth of his own work in comparison with that of the Montpellier physician.

Jean Astruc (1684–1766) was surely no less prominent than Vieussens or Hecquet. A student at Montpellier, where he took his M.D. in 1703, he was later named to the medical faculty of that school while serving at the same time as the professor of anatomy at Toulouse and First Physician to the King of Poland. Later he was to be named Professor of Medicine at Paris. It would seem that Astruc was fully aware of both iatrophysical and iatrochemical viewpoints. As early as 1702 he had published on the mechanism of fermentation and in 1710 he produced a *Dissertation Physico-Mathematique* on muscular motion that reflected the principles of Borelli.[66] But in the same year he began his work on digestion that was to result first in his *Mémoire sur la cause de la Digestion*, later reprinted by Hecquet in the 1712 edition of his work on digestion. In 1714 Astruc published his *Traité de la cause de la Digestion ou l'on*

[64] *Ibid.*, p. 509.

[65] *Ibid.*, vol. 2, p. 518.

[66] On Astruc, see the article by Pierre Huard in the *Dictionary of Scientific Biography*, vol. 1 (New York: Scribner's, 1970), pp. 322–324 with references. The early works referred to here are the *De motus fermentativi causa* (Montpellier, 1702); *Responsio critica animadversionibus F.R. Vieussens in tractatum de causa motus fermentativi* (Montpellier, 1702), and *Dissertatio physico medica de motu musculari* (Montpellier, 1710).

Portrait of Jean Astruc (1684–1766). Courtesy of the National Library of Medicine.

refute le nouveau Sistéme de la Trituration & du broïment. Et où l'on prouve que les alimens sont digerez & convertis en chile, par une veritable Fermentation.

In his treatise on digestion Astruc wrote first of the mechanistic interpretation of Pitcairne, whom he praised for his vivacious style. However, "we defend our views with moderation driven by a love of truth rather than passion."[67] And the truth clearly is that "all changes of food in digestion are products of fermentation" . . . and indeed . . . "we regard fermentation as the principle agent of the animal economy."[68]

In a short history, Astruc cited the ancients in support of fermentation[69] and disagreed with Hecquet's harsh view of van Helmont.[70] He affirmed that in digestion chemical dissolution occurs through the action of liquors called menstrua or ferments that penetrate and detach the parcels of food little by little until there is an entire dissolution. There are those who say that digestion only depends on trituration,[71] but in truth digestion is fermentation resulting from the leaven of the saliva, the bile, and the pancreatic juice. Those who, like Hecquet, insist on digestion through trituration are wrong on four counts. *First*, in digestion there is a veritable transmutation of the food, something that cannot be explained by mechanical means;[72] *second*, the force of the muscles of the stomach is not capable of crushing the food;[73] *third*, the structure of man's stomach proves that it is not destined to crush food;[74] and *fourth*, in the system of trituration one cannot explain hunger, indigestion, or appetite.[75] In short, Hecquet was wrong, digestion does result from fermentation.

The work of Astruc was replied to by a student of Pitcairne, Thomas Bower, a professor of mathematics and doctor of physic at the University of Aberdeen.[76] It is a work that Astruc replied to the following year.[77] Bower's

---

[67]Jean Astruc, *Traité de la cause de la Digestion ou l'on refute le nouveau Sistéme de la Trituration & du broïement. Et où l'on prouve que les alimens sont digerez & convertis en chile, par une veritable Fermentation* (Toulouse: Ant. Colomiez, 1714), sig. ã v^{v}.

[68]*Ibid.*, p. 6.

[69]*Ibid.*, pp. 12ff.

[70]*Ibid.*, pp. 30ff.

[71]*Ibid.*, pp. 52–53.

[72]*Ibid.*, p. 102.

[73]*Ibid.*, p. 138.

[74]*Ibid.*, p. 183.

[75]*Ibid.*, p. 209.

[76]I have been unable to locate a copy of the Bower (or Boër) response to Astruc. It is not in the holdings of the Wellcome Historical Medical Library or the British Library. Nor is it listed in the National Union Catalog.

[77]Jean Astruc, *Epistolae Joan. Astruc, quibus respondetur epistolari dissertatione Thomae Boeri de concoctione* (Toulouse, 1715).

work was originally to be included in the collection of Pitcairne's medical writings, but was eliminated by G. Sewell, the editor, who felt that the volume should only include genuine writings by Dr. Pitcairne. In any case, Pitcairne continued to publish in defense of his views until his death and he contributed a prefatory piece to the collected volume in which he wrote that "I will not call Astrucius' Book *Cacata Charta*, since Astrucius, in my Opinion seems never have gone to Stool, otherwise he must have perceived that the Muscles of the Abdomen have a Power of *Contraction* and *Expulsion*."[78]

It is not difficult to conclude from this early eighteenth-century debate that chemists continued to support their views in opposition to the mechanists. Joannis-Baptista Gastaldi questioned *An Alimentorum Coctio seu Digestio, à Fermentatione vel à Tritura fiat* (1713). After discussing the works of both Astruc and Vieussens to that date, he concluded that the coction of foods, or digestion, proceeds from fermentation rather than trituration.[79] The standard texts continued to be published well into the new century. The final edition of the *Opera* of van Helmont appeared in 1707,[80] while that of Thomas Willis saw several eighteenth-century editions to 1720 and that of Franciscus de la Boë Sylvius appeared as late as 1736. A four-volume edition of works by Vieussens was printed in 1774 and 1775. But the tide was clearly running in an opposite direction. Hecquet's work on digestion was published in an expanded two-volume edition in 1730 and again in 1747. In addition he attacked the chemists in his *Novus Medicinae Conspectus* (2 vols., 1722),[81] while in his posthumous *La Medecine naturelle* (2 vols., 1738), Hecquet complained that chemistry was a false analogy for the physician and that some young practitioners have been debauched by the charms of chemistry. Young doctors should accustom themselves to learn the causes of disease. They should seek the laws of health, which are the laws of motion, fluids, and solids. "They will learn nothing in the boutiques of the chemists."[82] Pitcairne's works reached a third edition by 1740 and Baglivi's opera had reached an eighteenth edition by 1737, while other editions appeared as late as 1788.

---

[78]Pitcairne, sig. B2 verso.

[79]Johannes Baptista Gastaldy, *Quaestio Medica. Proposita ab ill. ac nob. D.D. Joanne- Baptista Gastaldy, An Alimentorum Coctio seu Digestio, à Fermentatione vel à Tritura fiat* (Avenione: Joannem Delorme, 1713). After referring to both Astruc and Vieussens, Gastaldi concludes (p. 45) that "Ergo Alimentorum coctio, seu digestio fermentationi potius quam triturae, est tribuenda."

[80]See the lengthy and favorable review of the 1707 edition of the *Opera* of van Helmont in the 1708 volume of the *Journal des Sçavans*, pp. 123–129.

[81]*Journal des Sçavans* (1723), pp. 409–427.

[82]*Journal des Sçavans* (1739), p. 234.

A SPANISH DIGRESSION[83]

Although Spain would seem to be an unlikely country to facilitate understanding of the chemical debates of the seventeenth and eighteenth centuries, in fact it presents a very special case of more than ordinary interest.[84] The Spanish reaction to the work of Paracelsus and the introduction of chemical medicine is unique largely because of the impact of the Reformation.

Philip II's concern over the spread of the Protestant heresy led to draconian measures aimed at preventing the introduction of foreign ideas early in his reign.[85] All books were to be examined and licensed before publication. Those who imported books or published or circulated them without proper license were subject to death and confiscation of goods. The Spanish *Index* of prohibited books was published in 1559 and a series of auto-da-fés were held to stamp out religious heresy. As for Spanish students, they were ordered home from foreign universities within three months, except those at the theologically orthodox colleges in Bologna, Rome, Naples, and Coimbra.

The effectiveness of these measures is currently the subject of debate, but there is no doubt that there was a severe decline in Spanish science and medicine in the century from 1560 to 1660. Mid-seventeenth century visitors to Spanish universities were astonished by the low state of Spanish learning. Francis Willughby (1664) reported that the students at Valencia had not heard of the new philosophy and Lorenzo Magliotti (1668) wrote that med-

[83]The present section derives in part from two papers presented earlier: Allen G. Debus, "Chemistry and Iatrochemistry in Early Eighteenth-Century Portugal: A Spanish Connection," *Historia e Desenvolvimento da Ciência em Portugal: I Colóquio - até ao Século XX. Lisboa, 15 a 19 Abril de 1985*, Secretary-General of the Colloquium, Prof. Doutor António Vasconcellos Marques, 2 vols. (Lisboa: Academia das Ciências de Lisboa, 1986), vol. 2, pp. 1245–1262; Allen G. Debus, "Paracelsus and the Delayed Scientific Revolution in Spain: A Legacy of Philip II," *Reading the Book of Nature: The Other Side of the Scientific Revolution*, edited by Allen G. Debus and Michael T. Walton, volume XLI, Sixteenth Century Essays and Studies (Thomas Jefferson University Press, 1997), pp. 139–152.

[84]A letter from Professor James S. Amelang of the Universidad Autonoma de Madrid (2 March 1997) suggests that "there is an emerging point of view–it would be premature to call it a consensus–that locates the causes of this growing (and increasingly visible) cultural and scientific backwardness less in the repressive policies of the monarchy, and the Inquisition in particular, than in the deepening poverty of Castile, in economic, social and political terms."

[85]Henry Kamen, *Spain 1469–1714: A Society of Conflict* (London and New York: Longman, 1983), pp. 118–120.

ical teaching in Spain remained wedded to Galen. He added that anatomy had not been taught at Alcalá in a decade.[86] There was, in fact, a decline of teaching in all scientific areas.[87]

The situation inevitably affected the reception of Paracelsian thought in Spain. In most of the countries of western Europe, Paracelsian chemical medicine did not gain wide currency until the 1570s, well after Philip's decrees against foreign influence. Furthermore, the Paracelsian world view was set in a mystical-religious context of concern to the Church. It was not long before the name of Paracelsus began to appear in the Spanish *Index* (1583, 1584). In the 1632 edition he was classed as a Lutheran and most of his work was forbidden.[88] He was placed in the "first class of authors of damned memory" and we need only be reminded of the action of the Spanish Inquisition against van Helmont to realize the serious consequences of being considered a Paracelsian.[89]

With the attempt to prevent the importation of suspected foreign literature, the decline of the universities, and an inherent resistance to innovation, it is not surprising that there were few Spanish followers of Paracelsus in this period. Llorenç Coçar became professor of chemical medicine at Valencia in 1591 and produced a short work on the subject (1589) attacking the ancients and praising Paracelsus, but there is no reference to him after 1592, and his book, the *Dialogus veros medicinae fontes indicans*, survives only in a single copy.[90] There is also a work on practical chemistry (1598) by Diego Santiago

[86]Henry Kamen, *Spain in the Later Seventeenth Century 1665–1714* (London and New York: Longman, 1983), pp. 313, 319.

[87]José María López Piñero, *La Introducción de la Ciencia Moderna en España* (Barcelona: Ediciones Ariel, 1969), pp. 38–44.

[88]The problems of Paracelsus and the Spanish *Index* are discussed in anon., "La Historia de la Ciencia en España como Realidad Marginal en su Organización y Contexto Social," *Anthropos*, no. 20 (1982), 2–15 (4). See also J. M. Bujanda, *Index de L'Inquisition Espagnole (1583, 1584)* (Quebec: Centre d'Études de la Renaissance: Éditions de l'Université de Sherbrooke/Librairie Droz, 1993), pp. 546–547 (1583), 868–869 (1584).

[89]Debus, *Chemical Philosophy*, *2*, pp. 306–311.

[90]López Piñero has touched on the work of Coçar in many of his publications. He has also reprinted the *Dialogus veros medicinae fontes indicans in El "Dialogus" (1589) del Paracelsista Llorenç Coçar y la Cátedra de Medicamentos Químicos de la Universidad de Valencia (1591)* (Valencia: Cátedra e Instituto de Historia de la Medicina, 1977). Here he has added a lengthy introduction on Coçar and his work. In his *Clásicos Médicos Valencianos del Siglo XVI* (Valencia: Generalitat Valenciana, 1990), López Piñero has translated a lengthy passage from the *Dialogus*, pp. 129–134.

in which Paracelsus is referred to as an authority.[91] However, in general, the few references to Paracelsus are highly critical.[92]

This situation was not to change markedly until after the accession of Charles II in 1665. Although decisions during the regency were principally made by Queen Mariana, Don Juan José, the natural son of the late King Philip IV, exercised increasing power until his death in 1679.[93] The influence of Don Juan was important for the acceptance of the new science in Spain. Having an interest in all aspects of science and technology, we find his influence on the rapidly growing interest in these areas in the closing decades of the century.

The result of this is particularly interesting in chemical medicine. Texts that had been forbidden or at least imperfectly known earlier now became widely available. Thus, sixteenth-century Paracelsian texts–many of which were considered outdated in other countries of western Europe–were now being read along with the very recent iatrochemical works of Thomas Willis and Franciscus de la Boë Sylvius. Many Spanish physicians of the last two decades of the seventeenth century were now relatively unconcerned with the warning to be found in the *Index*. Rather, they saw in iatrochemistry a new observational and experimental basis for medicine.

Don Juan supported these medical chemists who rejected scholastic traditions, accepted the Harveyan circulation, and referred with interest to the scientific methodology of Francis Bacon and René Descartes. Don Juan appointed an Italian, Giambattista Giovanni (1636–1691), as his surgeon. Giovanni adopted a Spanish name, Juan Bautista Juanini, and he became the first true Spanish defender of chemical medicine.[94] Juanini turned first to an environmental problem, the polluted air of Madrid, which he blamed as the source of epidemics. The cause, he wrote, was due to dangerous salts in the air that had been released from the putrefaction of refuse in the streets

[91]Diego de Santiago, Destilador de su Magistad, Vezino de Sevilla, *Arte Separatoria y Modo de Apartar Todos los Licores, que se sacan por via de Distilacion: para que las Medicinas obren conmayor virtud, y prestiza* (Seville: Francisco Perez, 1598). Part 2, fol. 21 verso.

[92]López Piñero, *Ciencia y Técnica en la Sociedad Española de los Siglos XVI y XVII* (Barcelona: Editorial Labor, 1979), p. 378. Francisco de Quevedo, *Sueños y Discursos*, 2 vols. (Madrid: Editorial Castalia, 1993), p. 186.

[93]On the importance of Don Juan, see Kamen, *Spain in the Late Seventeenth Century*, pp. 320, 323; López Piñero, *Introducción*, p. 37.

[94]*Ibid.*, pp. 64–76.

through fermentation.[95] The plague he blamed on the malignant vapors rising from cadavers that infect the air and fill it with saline and poisonous atoms. This air, being carried by inspiration into the vessels of the lungs and heart, coagulates the blood and causes sudden death.[96]

Juanini rejected the academic science of the scholastics and turned rather to demonstrations and mechanical experiments. Like other Spanish chemists of the period, he enthusiastically supported the Harveyan circulation of the blood and explained respiration in terms of nitro-aerial particles. How sad it is, he remarked, that although the circulation of the blood is accepted throughout Europe, that it is not yet received in the medical schools of Spain.[97]

Even more influential was the *Carta filosófica, medico-chymica* (1687) of Juan de Cabriada, a son of a professor of medicine at Valencia. As a young man little more than twenty, Cabriada had been involved in a debate during a consultation with Galenists at Court.[98] This led to his writing the *Carta filosófica*, in which he refuted the authority of the ancients and insisted that observation and experience be the basis of authority in the study of nature. We stand on the shoulders of giants and we should turn to their work: Paracelsus, van Helmont, Sylvius, and Willis as well as Sanctorius, Descartes, Redi, Sydenham, Boyle, and Harvey, rather than Aristotle, Galen, and their followers.[99]

> How sad and shameful it is that, like savages, we have to be the last to receive the innovations and knowledge that the rest of Europe already has. And when we bring this to the attention of Spaniards who should know it, they get offended and are irritated by the truth. How true it is that to try and shift people from an established opinion is the most difficult thing to attempt.

---

[95]Jean-Baptiste Juanini, *Dissertation Physique, ou l'on Montre les mouvements de la fermentation: Les effets des matieres nitreuses dans les corps sublunaires, & les causes que alterent la pureté de l'air de Madrid*, trans. from Spanish to French by Jean-Joseph Courtial (Toulouse: D. Desclasson, 1685), pp. 2–9.

[96]*Ibid.*, pp. 36–40, 55.

[97]*Ibid.*, p. 93.

[98]Doctor Juan de Cabriada, *De Los Tiempos y Experiencias el Mejor Remedio al mal. Por la Nova-Antigua Medicina Carta Philosophica Medica Chymica* (Madrid: Lucas Antonio de Bedmar y Baldwin, 1687). López Piñero, *Introducción*, p. 76.

[99]*Ibid.*, p. 104.

Cabriada asked why the King did not form a Royal Academy like those that existed in France and England for the study of natural things through the promotion of physico-chemical experiments. And our eyes should be open to the wonders that the modern writers, new Columbuses and Pizarros, have discovered by means of their experiments, both in the macrocosm and the microcosm.[100] It is clear that Cabriada sought a total reform of Spanish science, but he was a physician and his interests were centered on a new medicine based on the Harveyan circulation and chemical medicine.

The *Carta filosófica* initiated a decade of debate. Some sought to hold firmly to tradition, while others supported Cabriada's call for reform. Christoval Tixedas (1688) wrote that Spaniards had always opposed sects and that this should hold true no less in medicine than in religion.[101] Chemical medicines may have some value, but chemistry must not be viewed as the foundation of medicine.[102] Spain had remained true to divine law and the Catholic faith without accepting other sects.[103] We should stand with Hippocratic and Galenic medicine as well as with Sacred Theology.

In 1690 José Gazola wrote an *Entusiasmos medicos* in which he supported Cabriada and attacked the Galenists. The same year Diego Mateo Zapata wrote his *Verdadera Apologia en Defensa de la Medicina Racional Philosophica* in which he held to much of traditional medicine, but distinguished between the new chemists and others who follow a diabolical path "as is seen in the infamous heresy of the chemists Paracelsus and his followers, whose infamous writings have justly been prohibited by the very just decision of the Holy Inquisition against all those who follow, aid and profess these false and diabolical doctrines as stated in the Expurgation of 1640."[104]

Clearly by the final decade of the century, Spanish readers were aware of many of the Paracelsian authors who had been denied to them in the *Index* of forbidden books. And if there was not yet any agreement regarding the value of medicinal chemistry, the call for a royal academy by Cabriada bore

---

[100]López Piñero, *Introducción de la Ciencia Moderna*, pp. 101–107.

[101]Christoval Tixedas, *Verdad Defendida, Respuesta de Fileatro, a la Carta Medico-Chymica, que contra los Medicos de la Iunta, de la Corte, y contra todos los Galenicos, le escriviò el Doctor Medico-Chymico D. Iuan Cabriada* (Barcelona: En casa de Antonio Ferrer y Balthasar Ferrer, 1688), fol.§§ 2^{r}.

[102]*Ibid.*, p. 46.

[103]*Ibid.*, p. 50.

[104]Diego Mateo Zapata, *Verdadera Apologia en Defensa de la Medicina Racional Philosophica* . . . (Madrid: Antonio de Zafra, 1690), p. 46. On Zapata, see José Guillhermo Merck Luengo, *La Quimiatria en España* (Madrid: Instituto "Arnaldo de Vilanova," 1959).

fruit in the founding of the Regia Sociedad de Medicina y otras Ciencias in Seville in 1700.

A second Spanish debate resulted from the *Hippocrates defendido* written by Dr. Miguel Marcelino Boix y Moliner (b.c. 1633) published in 1711.[105] In a long section he pictured a scene in which physicians from each major sect sought to cure a patient. The judge was Hippocrates and the physicians came to him one by one according to the antiquity of their sect. First came the Galenist, who suggested bloodletting as a cure.[106] He was dismissed curtly. Next came a Paracelsian accompanied by an Helmontian.[107] The latter served as spokesman and stated that chemistry had been undervalued in medicine. This infuriated Hippocrates, who said that he had spent much time on it and its philosophy. However, after all his study he remained convinced that diet is more effective than the nonsense of the chemists' arcanas, corallines, and alkahests. The Helmontian replied that Hippocrates could not treat "incurable" diseases without the chemists' secret arcanas. Angered by this, Hippocrates replied that van Helmont himself had been killed by one of his remedies. He refused to let the Helmontian reply because chemical physicians had not taken into account the days or times of illnesses and without this it would be impossible to cure an acute fever.

Next came a follower of Thomas Willis, but he was dismissed because the reasoning of Willis was too subtle for a practical physician.[108] As for Franciscus de la Boë Sylvius, there was little doubt that he was a successful doctor, but he did not have the answer for this illness either.[109] Descartes was turned away because his presentation did not distinguish between medicine and philosophy.[110] Then Baglivi approached Hippocrates and he presented his mechanical system of medicine, in which the body could be viewed as a

---

[105]Doct. D. Miguel Marcelino Boix y Moliner, *Hippocrates Defendido, de las Imposturas, y Calumnias, que algunos Medicos poco cautos le imputan: En particular en la Curacion de las enfermedades agudas: pues hasta aora todavia se ignora como las curava; con sola la Exposicion, ò comento del primer Aphorismo: Vita brevis, Ars vero longa, &c.* (Madrid: Matheo Blanco, 1711). I have used the Mexican edition, *Hipócrates Defendido* (Mexico, D.F.: Imp. Del Sagrado Corazón de Jesús, 1893). Here the "Junta de Médicos appears on pp. 236–262. On Boix y Moliner, see José F. Prieto Aguirre, "La Obra de Boix y Moliner. Historia de una Polemica," *Estudios de Historia de la Medicina Española* [Nueva Series I (1959–1960), no. 6 (441 pp. plus illustrations)].

[106]*Hipócrates Defendido* (1893), p. 238.

[107]*Ibid.*, p. 239.

[108]*Ibid.*, pp. 240–241.

[109]*Ibid.*, p. 242.

[110]*Ibid.*, pp. 243–244.

clock or an hydraulic machine.[111] He also rejected the consideration of times and days in illness as unimportant, thereby infuriating Hippocrates, who branded him a "taranbula physician"[112] who had incorporated into his system bad theory and practice and had put all medicine into confusion when he had entangled it with mathematics.[113]

Finally all of the sectarians went with Hippocrates to the bed of the sick person . . . that is, all except the Cartesian participant who did not have time to wait. At that time it became apparent that the only correct procedure would be to follow the method of Hippocrates.[114]

As in the case of Cabriada, the *Hippocrates defendido* led to an extended debate. Antonio Diaz del Castillo criticized Boix y Moliner in an *Hippocrates desagraviado* (1713) and an *Hippocrates Entendido* (1719) and Boix replied with an *Hippocrates aclarado* (1716). Dr. Don Antonio Alvarez del Corral prepared an *Hippocrates vindicato* (1713), in which he argued that chemistry was essential for the physician.[115] Not only does it teach us to analyze mixed bodies and to separate the pure from the impure,[116] it also has contributed to physiology.[117] Willis has discussed digestion in terms of fermentation and we have learned much of the properties and parts of the blood through its distillation. The bodily organs themselves have been compared to chemical vessels. In truth, even the anatomist must learn to understand chemistry.[118]

Also resulting from this debate was *La Pharmacopea Trivunfante de las Calvmnias y imposturas, que en el Hipocrates Defendido ha publicado el Doctor Don Miguel Boix* (1713) written by Felix Palacios. A chemist himself, Palacios defended van Helmont, Willis, and Sylvius,[119] while relying heavily on the

---

[111] *Ibid.*, pp. 250–252.

[112] Baglivi was well known for his study of the effects of the bite of the tarantula.

[113] *Ibid.*, p. 259.

[114] *Ibid.*, p. 260.

[115] Dr. Don Antonio Alvarez de Corral, *Hippocrates Vindicato y Reflexiones Medicas, sobre el Hippocrates Defendido* (Madrid: Por la Viuda de Juan Garcia Infançon, 1713), pp. 644–647.

[116] *Ibid.*, pp. 583–589, 643.

[117] *Ibid.*, pp. 644–647.

[118] *Ibid.*, p. 648.

[119] Felix Palacios, *La Pharmacopea Trivnfante de Los Calvmnias y imposturas, que en el Hipocrates Defendido ha publicado el Doctor Don Miguel Boix, Medico honorario, y de la Regia Sociedad Medica de Sevilla* (Madrid: Francisco Martinez Abad, 1713), pp. 35–36. After discussing Boix's attack on the chemists, Palacios went on to attack Boix for his views on Descartes, Gassendi, the atomists, Baglivi, and the mechanical philosophy, pp. 38–66 and *passim*.

work of Robert Boyle.[120] Palacios had earlier written his *Palestra Pharmacevtica, Chymico-Galenica* . . . (1706), a major Spanish pharmacopoeia that went through seven editions by 1792. Here Palacios concentrated on the preparation of specific medicaments, but the work could also be described as a chemical textbook because he defined terms and discussed chemical procedures.

A final Spanish example may be seen in Francisco Suarez de Rivera (c.1680–1754).[121] After taking his M.D. degree at Salamanca (1711), he had a distinguished career practicing in many of the major cities of Spain prior to moving to Madrid, where he became a physician to the nobility and the royal family. Suarez was very productive; he wrote more than forty books between 1718 and 1751, comprising some 15,000 printed pages. These books are all medically oriented. They range over all aspects of the medical sciences and they show his acquaintance with all medical sects.

He is of interest here primarily because of his views on the relation of chemistry and medicine. As for Paracelsus, Suarez wrote that it was of no concern to him whether or not he had been superstitious and had employed remedies invented by the Cacodemon or that he had learned from the books of black magic. These were matters for the Sacred Tribunal of the Inquisition. For Suarez the main concern was simply whether or not the work of Paracelsus was useful for the public health.[122]

And again, much as he was influenced by Galenic medicine himself, Suarez was convinced that chemistry was essential for the physician. The circulation of the blood is an imitation of a chemical circulation in a flask[123] and there is a true chemical surgery that proceeds without a knife.[124] Thus, the

---

[120]As an example, see his reference to the *Sceptical Chymist*, p. 70.

[121]On the life of Suarez de Rivera, see the short biography (with references) by J. M. López Piñero in the *Diccionario histórico de la ciencia moderna en España*, *2*, pp. 339–341 and the two monographs by Luis S. Granjel, *Francisco Suarez de Rivera: Médico salamantino del siglo XVIII* (Salamanca: University of Salamanca Press: Cuadernos de Historia de la Medicina Española, Monografias IV, 1967) and José Luis Valverde, *La Farmacia y las Ciencias Farmaceuticas en la Obra de Suarez de Rivera* (Salamanca: University of Salamanca Press: Cuadernos de Historia de la Medicina Española, Monografias XIII, 1970).

[122]Suarez de Ribera, *Anatomica Chymica, Inviolable, y Memorable* (Madrid: Manuel de Moya, 1743), pp. 21–22.

[123]Suarez de Ribera, *Medicina Invencible Legal, O Theatro de Fiebres Intermitentes Complicadas* (Madrid: Francisco del Hierro, 1726), pp. 262–263. On page 298 he goes on to give a lengthy discussion of a Dr. Botoni on the circulation of the blood.

[124]Suarez de Ribera, *Canones Particulares de Cirugia, con que se Libertan Muchos Desahuciados, si al Sagrado de sus fuentes se Refugian* (Madrid: Manuel de Moya, 1751), p. 73; *Cirugia Methodica Chymica Reformada* (Madrid: Francisco Laso, 1722), p. 8.

physician-chemist must know the three Paracelsian principles and the use of the chemical knife, fire, which separates pure from the impure and brings about new pharmaceutical preparations.[125]

Suarez still reflected the Renaissance interest in natural magic. "We rightly affirm that magic is the most subtle science of Physics which consists in the investigation of phenomena which we observe in natural as well as celestial bodies."[126] For Suarez, Paracelsus was the "grande Magico."[127] With this understood, we need hardly be surprised to find Suarez seeking the formula of van Helmont's alkahest,[128] Paracelsus' magnetic or sympathetic powder,[129] and the true preparation of *aurum potabile.* With Suarez de Rivera, surely one of the most prominent Spanish physicians of the first half of the eighteenth century, we still see an interest and concern for sixteenth- and early seventeenth-century topics. In regard to chemical medicine, it is as if he were living a century earlier.

CONCLUSION

The late seventeenth century shows a continued interest in Helmontian iatrochemistry through the work of Willis, Sylvius, and surely, the early writings of Robert Boyle. But at the same time the successes in the physical sciences culminating in the work of Isaac Newton resulted in a lessened interest in the texts of the ancient physicians and naturalists. Now there was a new hope that medicine, by being mathematicized, might find its own Newton and reach a similar level of certainty. The chemical philosophers, previously obsessed with their debates with the followers of the ancients, now found themselves opposed to the medical mathematicians and mechanists. The work of Sanc-

---

[125]F. Soares da Ribeyra, *Cirurgia Methodica e Chymica Reformada*, traduzida de Castelhano em Portuguez . . . Joseph Gomes Claro (Lisboa Occidental: Na Officina Ferreyrenciana, 1721), p. 128. Suarez de Ribera, *Anatomia Chymica*, pp. 19–27.

[126]Suarez de Ribera, *Amenidades de la Magia Chyrurgica y Medica, Natural* (Madrid: Fernandez de Arroyo, 1736), p. 38.

[127]*Ibid.*, p. 75.

[128]Suarez de Ribera, *Illustracion, y Publicacion de los diez y Siete Secretos del Doctor Juan Curvo Semmedo, confirmadas sus virtudes con maravillosas observaciones* (Madrid: Domingo Fernandez de Arroyo, 1732).

[129]Suarez de Ribera, *Escrutinio Medico ò Medicina Experimentada* [John Crerar Library copy missing title page. Introductory letter dated 1723], p. 271. *Amenidades de la Magia*, p. 73.

torius early in the century and that of Borelli in 1679–1680 offered considerable promise to those who sought a quantified and mechanized medicine. Both Baglivi in Italy and Pitcairne in Scotland presented far-ranging iatrophysical works before the turn of the century and although traditional iatrochemists continued to publish, the main conflict between the medical chemists and mechanists seems to have shifted to the interpretation of digestion by 1710. The topic was an old one–van Helmont had established the role of acid in the stomach–but the hope that medicine might benefit from a mathematical-mechanical approach encouraged not only Pitcairne, but also Baglivi and Hecquet, to present mechanistic interpretations. They were opposed by equally prominent iatrochemists such as Vieussens and Astruc in a debate that reached its peak in the second decade of the century. While one is accustomed to think of a Newtonian triumph in the physical sciences before the close of the seventeenth century, it is clear that the debate of the chemical physicians with the medical mechanists survived well into the next century.

The case of Spain presents an interesting corollary to the above. Shut off from the new developments in the sciences after 1560 because of Philip II's fear of the spread of the Protestant heresy and his insistence on an orthodox Catholicism, Spanish science and medicine gained little from the developments in other western European countries. However, this changed when Philip IV died in 1665 and his natural son, Don Juan, began to foster interests in all of the sciences and technology. At this time the texts of the past century were read and studied. Even Paracelsus was in large part new to Spanish scholars. His works were not widely known elsewhere until after 1560, so few of them had been available. Those that had been known in Spain were to be condemned by the Spanish Inquisition. However, in the closing decades of the seventeenth century his work seemed as current as that of Bacon, Descartes, Willis, and Newton. Indeed, we find a defense of Paracelsus in the work of Suarez de Rivera well into the eighteenth century.

There is a sense of "catch-up" in the Spanish literature of this period, as we have seen in the work of Juanini and Cabriada. Not only in the founding of the Royal Academy in Seville (1700), but also in the debate following the call for a new science in the *Carta filosófica* (1687) of Cabriada, and the debate after the publication of Boix y Moliner's *Hippocrates defendido* (1711), we see the rapid spread of the new science in Spain. To be sure, it had taken a different course, but some of the same problems that divided iatrophysicists and iatrochemists in other countries were now reflected among Spanish scholars.

CHAPTER 6

Chemistry and Medicine in the Early Eighteenth Century

The great modern Corrupter both of the Theory and Practice of Physick, was certainly Bellini; who seems not to have understood any one Principle of animal Motion, but constantly ascribes to one Law of Nature what really belongs to another . . . But before Bellini, the empyrical enthusiastick Chymists, such as Paracelsus, Van Helmont, and others, had introduced into Physick, the Principles of the specifick and corpuscularian Philosophy, and carried the Art of Healing beyond all Human Judgment and Comprehension; which Bellini's pretended mechanical Scheme did not at all mind, but rather made it worse than before.[1]

T. Morgan, *The Mechanical Practise of Physick* (1735)

The eighteenth century witnessed a change from the iatrochemical-iatrophysical debates of the previous century to the changes associated with Lavoisier in the final decades, generally referred to as the chemical revolution. Here we will be concerned primarily with the first half of the century, concentrating on the work of Boerhaave, Stahl, and their contemporaries. And here too we will have more interest in their views on the relation of chemistry to medicine than on the rise of the phlogiston theory.

[1]Thomas Morgan, M.D., *The Mechanical Practice of Physick: In which The Specifick Method is Examin'd and Exploded; and the Bellinian Hypothesis of Animal Secretion and Muscular Motion, Consider'd and Refuted. With some Occasional Remarks and Scholia on Dr. Lobb's Treatise on the Small Pox, Dr. Robinson on the Animal Oeconomy, and Professor Boerhaave's Account of the Animal Spirits and Muscular Motion* (London: T. Woodward, 1735), pp. xiii–xiv.

APOTHECARIES, PHYSICIANS, AND CHEMISTS

A legal conflict between the apothecaries and the fellows of the Royal College of Physicians regarding the practice of medicine by the former occurred in the early years of the new century. In 1701 William Rose, a London apothecary, was sued by the College for practicing medicine, which was forbidden by an Act of Henry VIII. Rose had in fact ministered and prepared medicines for one John Seale without a license or direction from the College or any of its fellows. Nor did he take any fee for his advice. The judgment was against Rose and in favor of the long-standing privileges of the College. However, since the apothecaries generally gave medical advice and prescribed medicines without the license of a physician, they moved to reverse the decision. In this they were successful (1704) and although they were not allowed to charge for their advice, they had established their right to practice. After this result the long-standing dispute between the physicians and apothecaries died down.[2]

The background to the Rose Case was discussed by Robert Pitt (1653–1713), a fellow of the Royal College of Physicians and the Royal Society who had taken his M.D. at Oxford in 1682. Later he became Professor of Anatomy at Oxford. Pitt objected to the claims of the chemists as well as the apothecaries. He explained that the physicians had been corrupted by the Fire of London. Many had left the City and while they were gone the apothecaries had taken over. When at last the physicians had returned, they were dependent on the apothecaries for referrals. They then found out that the apothecaries wanted to prescribe large quantities of expensive medicines.[3] "Are not, say they, The new *Hypotheses*, and Chemical Fancies preferable to all the Old Observations of the Actions and Passions, and Disorders of the Body, and their Remedies old and out of Fashion, when ours are new and but lately invented?"[4] Physicians used to diagnose the ills of their patients

[2]*A History of the Worshipful Society of Apothecaries of London*, volume I 1617–1815, abstracted and arranged from the Manuscripts and notes of the late Cecil Wall by the late H. Charles Cameron, Revised, annotated, and edited by E. Ashworth Underwood (London, New York and Toronto: Oxford University Press for The Wellcome Historical Medical Museum, 1963), pp. 132–135 and documents, pp. 389–402. On the Rose case see also Harold J. Cook, "The Rose Case Reconsidered: Physicians, Apothecaries and the Law in Augustan England," *Journal of the History of Medicine and Allied Sciences, 45* (1990), 527–555; Harold J. Cook, *The Decline of the Old Medical Regime in Stuart London* (Ithaca and London: Cornell University Press, 1986), pp. 246–249.

[3]Robert Pitt, *The Antidote: Or The Preservative of Health and Life, And the Restorative of Physick To Its Sincerity and Perfection* . . . (London: John Nutt, 1704), p. 98.

[4]*Ibid.*, p. 124.

and prepare the proper medicines themselves and they should return to this practice.

Pitt was particularly alarmed by the claims of the chemists who killed with their prescriptions of gold, silver, copper, mercury, antimony, and arsenic. "What Numbers of the Sick have been burnt by the spirits, and Oyles, and Essences, and Elixirs, or chill'd to the loss of the Vital Heat by the Acids, and the innumerable combinations of them and their fixt Salts. The Magisteries and dead Earths have destroy'd as many by being unactive, and effecting nothing to the Preservation and Recovery of the Diseas'd."[5] But although Pitt felt that many of the chemical preparations had already been discarded by physicians, "Chymistry continues still to assault and depress the more Rational and useful Physick, by intruding very many impertinent *Hypotheses*, on the Art, which can admit of none."[6] Of particular concern to Pitt was the application of key elements of the mechanical philosophy to medicine.

> Give me leave to suppose, says the Philosopher, That Atoms are the Principles of all Things, and that they are of the very Figures and Magnitudes, which shall be appointed by me, and I will instantly describe all the Formations and Operations of all Corporeal Beings. Allow me, says the Physician, That the Spirits, the Blood and the Humours, are made of the figur'd Particles, and move in the manners I have concerted, there shall Remedies be applied to them, of these and these Figures and Particles, which shall soon reduce all their disorderly Motions. . . . Let the Mathematicks be allow'd to describe Fevers and Convulsions, the Medicines shall be applied to Cure them Mathematically.[7]

It would be much better, he thought, to follow the ancient authors and their followers today who depend on their senses "and the necessary and certain Deductions from clear and undoubted Observations, the Causes and Symptoms of Diseases, and the Qualities of Medicines and their constant Operations." How could this true and useful medicine survive when practitioners are "perverted by the Fame of Chymistry, and the new *Hypotheses* . . . ?"[8]

---

[5] *Ibid.*, p. 111.

[6] *Ibid.*, p. 112.

[7] *Ibid.*, p. 113.

[8] *Ibid.*, pp. 113–114.

Pitt's distrust of the chemists was echoed thirty years later by Thomas Morgan (d. 1743), an M.D. and self-styled "Christian deist" who served as independent minister of Burton, then Marlborough, before being dismissed for heterodoxy. He complained of the followers of Paracelsus as well as those of the mechanists. The views of the former and of his latter-day disciple, van Helmont, had pushed medicine "beyond all human Judgment and Comprehension." But Bellini and his mechanists made the situation worse. Indeed, Bellini should be considered the "great modern Corrupter both of the Theory and Practice of Physick."[9] In contrast to both extremes, Morgan praised Thomas Willis, to whom he ascribed a true understanding of the chemistry of the body.[10]

Morgan was particularly incensed by Bryan Robinson's (1680–1754) *Treatise of the Animal Oeconomy* (1732). A professor of physic at Trinity College, Dublin, and later three-time president of King and Queen's College of Physicians, Robinson published both medical and mathematical works. Referring to the *Treatise of the Animal Oeconomy*, Morgan stated that "the Mathematical Part of this Book is nothing but a Heap of the grossest Contradictions, Absurdities, and the most obvious unpardonable blunders . . ."[11] In addition, many of the views relating to a chemical physiology of the body, particularly the acid-alkaline concept, were flawed because "the particular and specifick Qualities and Properties of acids and alkaline Medicine, are entirely alter'd and destroy'd by the digestive Action of the Stomack, before they can be convey'd to and mix'd with the blood."[12] Nor should the physician concern himself with fermentation. "I know that this Doctrine of animal Fermentation, and the different Kinds and Degrees of them, in the Ventricles of different Glands, has been absolutely rejected, and very much exploded by some modern mechanical or immechanical Writers; particularly by *Bellini* and his Followers. But then it is certain, that even *Bellini* himself knew nothing of the true Laws of Motions, and had no Regard at all to the Chymistry of Nature."[13]

Although Morgan was enough of a mechanist to explain the circulation of the blood in terms of tubes and pipes and to discuss the motion of the

---

[9]Morgan, *op. cit.*, pp. xiii–xiv.

[10]*Ibid.*

[11]*Ibid.*, p. 73.

[12]*Ibid.*, p. 135.

[13]*Ibid.*, p. 139.

blood in terms of velocity, pressure, and diameters of the vessels, he felt that a mechanical understanding of the body "consists in regulating the natural Secretions or Evacuations, by promoting such as are defective, and restraining such as are excessive and redundant."[14] And yet, with all of his concerns about the use of chemicals, Morgan could see no ill effects in prescribing to a patient an ounce or two of crude mercury metal daily for a period of up to six weeks![15]

CHEMISTRY IN HISTORY: FREIND AND LeCLERC

Historical arguments had been brought forth to support or to attack the claims of the chemists since the sixteenth century. In 1585 Rychard Bostok had asserted that the truths of the alchemists could be connected with the divine knowledge given to Adam and that while this knowledge had decayed in the following centuries, some remained with Plato and his followers. It was Aristotle who had turned against his mentor and who had spawned a host of disciples who still burdened the world through the educational system. But surely Hermes, the founder of true chemistry, lived long before Aristotle and could claim truths unknown to the Peripatetics. The alchemists of the Middle Ages preserved these truths and it was the glory of Paracelsus to restore alchemy to medicine as it had existed in the early days of mankind.[16]

The question of the antiquity of chemistry and its relation to medicine continued to vex scholars for well over a century and proved to be particularly difficult for physicians who approved of chemically prepared medicines, but who sought to discredit the role of Paracelsus in their introduction. A protracted debate occurred between Hermann Conringius (1606–1681), a classicist who taught at Helmstedt, and Olaus Borrichius (1626–1690), a

---

[14] *Ibid.*, p. 43.

[15] *Ibid.*, p. 206. Morgan was not the only one convinced of the value of metallic mercury. See also Thomas Harris, *A Treatise on the Force and Energy of Crude Mercury. Proving the usefulness and innocency of its internal application, but a great variety of experiments and histories of cases, acute and chronick* (London: E. Symon, 1734).

[16] R. Bostocke, Esq. (Rychard Bostok), *The difference betwene the auncient Phisicke . . . and the latter Phisicke* (London: Robert Walley, 1585). The historical chapters of this work have been reprinted with an introduction and annotations in Allen G. Debus, "An Elizabethan History of Medical Chemistry," *Annals of Science, 18* (1962, published 1964), 1–29.

professor of chemistry and botany in Copenhagen. Conringius was an Aristotelian and a Galenist who opposed the chemical medicines of Paracelsus and rejected the historical existence of Hermes Trismegistus (1648). Twenty years later Borrichius replied in the *De ortu et progressu chemiae*, in which he sought to prove the very great antiquity of chemistry. There seemed little doubt to him that Tubal Cain was aware of chemical processes and as for Hermes, Borrichius stated that "Verum quo auctore coeperit primum in Aegypto *Chemia*, opportunè aliquis percunctabitur. Equidem nil dubitamus *Hermeti* sive *Mercurio* tum Medicinae, tum Chemiae adscribere originem . . ."[17] In short, Conringius was mistaken regarding the antiquity of the art and the role of Hermes Trismegistus in its transmission to the rest of the world from Egypt. Conringius replied to Borrichius in a newly revised edition of his 1648 work on hermetic medicine. Here he attacked hermetic knowledge as a whole, which he assured the reader had been so corrupted by time that we hardly know what the original texts meant. As for Paracelsus and his followers, he was adamant. Their work had corrupted all philosophy. Their three principles were useless and their metallic medicines had been plagiarized from earlier physicians such as Arnald of Villanova and Raymond Lull. As for the rest, it is "partim impiam . . . partim vanam et absurdum."[18] As one might expect, Borrichius could not let this stand unanswered. In 1674 he published yet another book defending the antiquity and truth of Egyptian knowledge and insisted on an historical Hermes who had discovered how to transmute the base metals to gold. The great Greek authorities had studied in Egypt and it was to the glory of Paracelsus that he had rediscovered the truths known to the Egyptian adepts.

The exchange between Conringius and Borrichius was given wide publicity through lengthy reviews in the *Philosophical Transactions* of the Royal Society of London in 1668 and 1674, and in the *Journal des Sçavans* in 1675.[19]

---

[17]Olaus Borrichius, *De ortu et progressu chemiae dissertatio* as printed in the *Bibliotheca chemica curiosa*, ed. Jean Jacques Manget (2 vols., Geneva: Chouet, G. De Tournes, et al., 1702), vol. l, pp. 1–37 (12). On the debate between Conringius and Borrichius, see also Allen G. Debus, "History With a Purpose: The Fate of Paracelsus," *Pharmacy in History*, *26* (1984), 83–96.

[18]Hermann Conringius, *De Hermetica Medicina Libri Duo* . . . (2nd edition, Helmstedt: typis & sumptibus H. Müller, 1669), p. 345.

[19]See the reviews of Olaus Borrichius, *Hermetis, Aegyptiorum, et chemicorum sapientia ab Hermanni Conringi animadversionibus vindicata per Olaum Borrichium* (Hafniae: sumptibus P. Hanboldi, 1674) in the *Philosophical Transactions of the Royal Society of London*, *10* (no. 113)(April 26, 1675), 296–301 and Borrichius, *De ortu et progressu* in the *Philosophical Transactions of the Royal Society of London*, *2* (no. 39)(1668), 779. See also the review of the *Hermetis Aegyptiorum et chemicorum* . . . in the *Journal des Sçavans*, *3* (1675), 209–211.

The points discussed regarding the antiquity of chemistry were considered of sufficient significance for the major texts, the *De Medicina Hermetica* of Conringius and the *De ortu et progressu chemiae* of Borrichius, to be considered fundamental sources for chemical historians at least until the close of the eighteenth century.

Nevertheless, as we have seen, the sciences were changing in the late seventeenth century. Those who sought truth in the mechanical philosophy let this affect their interpretations and we see this in the debate between Daniel LeClerc (1652–1722) and John Freind (1675–1728). LeClerc studied medicine at Paris and Montpellier prior to taking his M.D. at Valence. He published a *Bibliotheca Anatomica* (1688) and a *Chirurgie complète* (1695) before publishing his *Histoire de la médecine* (1696). Although the *Histoire* only covered the development of medicine to the time of Galen, LeClerc surveyed earlier histories, and when writing of Egyptian contributions, stated that "Hermes, or Mercury, or Thoth . . . by some confounded with Moses," was "the Inventor of Physick."[20] Nevertheless, there was little in this edition on the question of chemical medicine.

In 1723 LeClerc brought forth an expanded version of his *Histoire* that included a fifty-five page "Essai d'un Plan pour servir à la Continuation de l'Histoire de la Médecine Dépuis la fin du siècle second de J.C. jusqu'au milieu du dix-septième." Here the author's interest in chemistry and chemical medicine is evident. He discussed the ancient Greek alchemists as well as the chemical sections of Dioscorides before going on to the Arabic authors. Avicenna was of particular importance here since LeClerc felt that he was the first to introduce chemistry to medicine. There seemed little question that the Arabs were the first to introduce distilled waters or that they were well acquainted with the earlier Greek alchemical texts. From Islam chemical knowledge shifted to the West and LeClerc discussed Roger Bacon, John and Isaac Hollandus, Jean of Rupescissa, Basil Valentine, and Arnald of Villanova, whom he considered "a great chemist" who had made gold.[21]

[20]Daniel LeClerc, *The History of Physick, Or An Account of the Rise and Progress of the Art, And The Several Discoveries therein from Age to Age*, trans. Dr. Drake and Dr. Boden. With Additional Notes (London: D. Brown, A. Rope and J. Leigh, 1699), p. 11.

[21]Daniel LeClerc, *Histoire de la Médecine, Où l'on voit l'Origine & les Progrès de cet Art, de Siècle en Siècle; les Sectes, qui s'y sont formées; les noms des Médecins, leurs découvertes, leurs opinions, & les circonstances les plus remarquables de leur vie* (Nouvelle Edition, Amsterdam: Aux Depens de la Compagnie, 1723), pp. 785–790. I have also used the edition of 1729 with the same pagination.

At this point LeClerc turned to Paracelsus "who sought to establish an entirely new medicine on the ruins of the old."[22] The account of Paracelsus is not altogether complimentary, because he quoted extensively from Thomas Erastus on the religious views of Paracelsus and discussed his beliefs regarding sylphs, gnomes, and the homunculus and even cited Conrad Gesner to the effect that Paracelsus had a personal demon. He added that Paracelsus had introduced the three principles and attacked the traditional humors. He had coined new words and gave new meanings to old words such as *anatomia*, *astrum*, *manna*, *essentia*, and *ens.*[23] All of this made the writings of Paracelsus difficult to understand.

Proceeding from these criticisms to a survey of the Paracelsian system, LeClerc discussed the macrocosm-microcosm analogy as used in the works of the Swiss-German author and then went on to his understanding of quintessences and belief in a universal remedy. LeClerc understood Paracelsian disease theory in terms of an application of the three principles, salt, sulphur, and mercury. He also explained the theory of signatures and the emphasis on medicinals produced chemically from metals such as potable gold, antimony compounds, and mercury precipitate.[24] LeClerc noted that the surgical work of Paracelsus treated wounds and ulcers primarily and that he used herbal remedies on occasion as well as chemicals. Therefore, although the reader might benefit from the medical value of the work of Paracelsus, he would have to avoid his sections on astrology, geomancy, chiromancy, and the kabbala. Indeed, his medicine could be said to be joined with magic and LeClerc specifically compared Paracelsus with his contemporary, Henricus Cornelius Agrippa.[25] In short, LeClerc presented to the reader a balanced assessment of Paracelsus, pointing to the valuable as well as the superstitious aspects of his work.

In England John Freind had looked forward with impatience to the new edition of LeClerc's history. An Oxford M.D. (1703), Freind became Reader of Chemistry there (1704) and was later to become physician to King George I and Queen Caroline, and Fellow of both the Royal Society (1712) and the Royal College of Physicians (1716). As a member of Parliament he became implicated in Francis Atterbury's Jacobite plot and was imprisoned in the Tower of London for a time.

[22] *Ibid.* (1729), p. 792.

[23] *Ibid.* (1729), pp. 795–805.

[24] *Ibid.* (1729), pp. 806–812.

[25] *Ibid.* (1729), pp. 812–819.

Both a prominent physician and chemist, Freind early established himself as an outspoken Newtonian.[26] In his chemical work, Freind openly attempted to divorce himself from earlier chemical authors. He sought to explain chemical reactions as the interaction of spherical atoms possessed of forces similar to the gravitational forces that Newton had postulated for the solar system. He postulated an attractive force between these particles that decreased with distance faster than the inverse square law. His was an attempt to establish a Newtonian chemistry as part of a greater Newtonian system.

But it is in Freind's history of medicine that we find the material we seek. He had anticipated eagerly the expanded edition of LeClerc's *Histoire* because of his high opinion of the earlier work. "We find there very amply and clearly represented all the Philosophy, the Theory, and the Practice of the ancient Physicians; so that there is scarce a Notion, a Distemper, a Medicine, or even the name of an Author, to be met with among them, of which he has not given a full and exact account."[27] But in the new edition his *Plan* that continues the work to the mid-sixteenth century is much too short and suffers greatly by "relating all the obscure jargon and nonsense of that illiterate Enthusiast, *Paracelsus.*"[28] Accordingly, Freind planned his work to deal with the period from Galen to the beginning of the sixteenth century . . . thus giving a more extensive account of the period covered in LeClerc's *Plan*, but avoiding a detailed treatment of Paracelsus. Indeed, the sixteenth century was notable in medicine for many developments that would have been far more important for LeClerc to have discussed than "*Paracelsus's* idle system."[29] He, for one, would not spend time on it.

And yet, Freind was a chemist who was interested not only in Newtonian mechanics, but in the use of chemicals in medicine. But how might this advance be discussed without giving credit to Paracelsus? The answer was to

[26]On Freind, see J. R. Partington, *A History of Chemistry, 2* (London: Macmillan, 1961), pp. 479–482 and Arnold Thackray, *Atoms and Powers: An Essay on Newtonian Matter-Theory and the Development of Chemistry* (Cambridge: Harvard University Press, 1970), *passim.* His chemical lectures, the *Praelectiones Chymicae* (1709), reflect Newtonian mechanism rather than traditional medical chemistry. I have used the Naples edition of 1722, which is appended to Hadrian Mynsicht's *Thesaurus et Armamentarium Medico-Chemicum.*

[27]John Freind, M.D., *The History of Physick; From the Time of Galen, to the beginning of the Sixteenth Century. Chiefly with Regard to Practice. In a Discourse written to Doctor Mead* (Fourth edition, 2 volumes, London: J. Walthoe, 1744, 1750; reprinted New York: AMS Press, 1973), *1*, pp. 1–2.

[28]*Ibid.*, *1*, p. 2.

[29]*Ibid.*, *2*, p. 336.

be found in the historical record. "I shall not here enter into any dispute concerning the Origin of medicinal Chymistry; only I must observe, that if it be, as perhaps it may be, derived from the *Arabians*, the honour of the invention ought rather to be restored to *Rhazes:* . . . This is the first Writer, I believe, who gives the least hint of *Chymical* Medicines; for what we find in the old *Greek Chymists*, as they are called, relates only to the fusion or transmutation of Metals."[30] True, LeClerc found the origin of chemical medicine in the distillation of rose water in the medical work of Avicenna, but Freind rejected this and opted rather for the earlier work of Rhazes, in which chemical operations are described in detail. Although Freind had little more to say about chemical medicine, it was in this fashion that he denied the importance of Paracelsus both by rejecting his claim to be the originator of chemical medicine and by casting scorn on his medical system as a whole.

HERMAN BOERHAAVE AND THE MECHANIZATION OF MEDICINE AT LEIDEN

There were few physicians or scholars in the eighteenth century who had a greater influence on their contemporaries than Herman Boerhaave.[31] Born the son of a minister of the Dutch Reformed Church on New Year's Day, 1669, Herman had an early interest in theology and philosophy and he enrolled at the University of Leiden at the age of fifteen. A brilliant student, Boerhaave held a series of public disputations before giving an academic oration on Cicero's interpretation of Epicurus (1689). This oration was well received and Boerhaave was presented with a gold medal for his effort. The following year he took his degree as a doctor of philosophy.

Boerhaave turned then to the study of divinity and in his free time he taught mathematics and worked in the library to support himself now that his scholarship had expired. In 1691, however, he turned his interests to medicine. At that time Leiden had a medical faculty, but Boerhaave later wrote that he did not attend any lectures and that he was largely self-taught in this

[30] *Ibid.*, *1*, pp. 278–279.

[31] The standard account of his life and work is G. A. Lindeboom, *Herman Boerhaave. The Man and His Work* (London: Methuen & Co., Ltd., 1968). Lindeboom has also summarized this work in his article on Boerhaave in the *Dictionary of Scientific Biography*, vol. 2 (New York: Scribner's, 1970), pp. 224–228. He has also prepared two volumes of *Boerhaave's Correspondence* (Leiden, 1962, 1964).

Portrait of Hermann Boerhaave (1669–1738). Courtesy of the National Library of Medicine.

field. He read Vesalius and Caspar Bartholin (1585–1629) on anatomy as well as the ancient Greek authors. Among modern authors he was attracted primarily by Thomas Sydenham (1624–1689), who had emphasized personal observations and experience and had had relatively little interest in theories. Simultaneously he became fascinated with chemistry through his acquaintance with David Stam, who was then preparing an edition of John Francis Vigani's *Medulla chemiae.* Stam had worked earlier with Franciscus Sylvius and it is clear that Boerhaave was acquainted not only with the work of this great Leiden iatrochemist, but also with that of Paracelsus, van Helmont, Tachenius, and the earlier alchemists. In addition to these studies, he read botanical works and examined animals through vivisection.

Considering the fact that he was largely self-taught in medicine, it may not be too surprising that Boerhaave chose not to seek his degree at Leiden. The professors in medicine there did not know him and they might well have felt offended that he had not attended their lectures. In any case, Boerhaave went to the less prestigious university at Harderwijk, where the fees were much less than at Leiden. It was a short process. He left Leiden on 11 July 1693 and matriculated at Harderwijk the following day. He was examined on the thirteenth on the whole of medicine, then discussed a case of apoplexy and explained an aphorism of Hippocrates. Having completed this and the defense of his thesis, he graduated on the fifteenth of July. The thesis itself, *De utilitate explorandorum in aegris excrementorum ut signorum*, referred frequently not only to Hippocrates, but also to Paracelsus and van Helmont. The defense completed, Boerhaave returned to Leiden the following day.

Back in Leiden, Boerhaave became a private physician, spending his free time in the study of mathematics, physics, anatomy, botany, and chemistry. In 1701 he was appointed lecturer in medicine at the University of Leiden for a period of three years. With the appointment Boerhaave gave an inaugural address. In it he spoke of the need of physicians to return to Hippocrates. Although this ancient Greek physician was clearly the most important of all medical writers, Boerhaave also told his audience not to ignore modern authors such as Vesalius, Harvey, and the Bartholins . . . and especially Thomas Sydenham, "that shining light of England, a Phoebus in our Art."[32] And although false chemists are to be avoided, "I love and study

[32]The major academic orations have been gathered and published in translation. *Boerhaave's Orations*, translated with introductions and notes by E. Kegel-Brinkgreve and A. M. Luyendijk-Elshout, published for the Sir Thomas Browne Institute (Leiden: E. J. Brill/Leiden University Press, 1983). Here see the "Oration to Recommend the Study of Hippocrates," p. 78.

chemistry, as it has now been purged by the diligence of the present age."[33] He referred specifically to the work of van Helmont, Tachenius, and Boyle.[34] Boerhaave's love of experience led him in this oration and throughout his life to denegrate Descartes and his followers, who preferred to begin with general principles rather than observation and experience.[35] His dislike of the Cartesians was to be a persistent theme in his writings.

In 1703 Boerhaave was invited by the Curators of the University at Groningen to a professorship. He declined and then informed the Curators at Leiden of the offer. Now aware of the growing importance of Boerhaave, they sought to promote him. There was no opening in the medical faculty, which was fixed at five professors, but he was promised a professorship when the next vacancy occurred. His salary was raised and he was to give a formal academic oration. This address, "On the Usefulness of the Mechanical Method in Medicine," was given on 24 September 1703. Here Boerhaave told his audience that although mechanics had proved most useful in physics, that it had generally been considered useless in medicine and consequently had been neglected.[36] He, on the other hand, was determined to show that "mechanics is most useful and absolutely necessary in medicine."[37] Indeed, Boerhaave felt that the human body is best understood in terms of mechanics.[38] In the lectures to his medical students, Boerhaave explained that

> The solid Parts of the human body are either membranous Pipes, or Vessels including the Fluids, or else *Instruments* made up of these, and more solid Fibres, so formed and connected, that each of them is capable of performing a particular Action by the Structure, whenever they shall be put in Motion; we find some of them resemble *Pillars, Props, Cross-Beams*, Fences, Coverings, some like *Axes, Wedges, Leavers*, and *Pullies*, others like *Cords, Presses* or *Bellows*; and others again like *Sieves, Strainers, Pipes, Conduits*, and *Receivers*; and the Faculty of performing various Motions by these Instruments, is called

[33] *Ibid.*, p. 79.

[34] *Ibid.*, p. 83.

[35] *Ibid.*, p. 80.

[36] "Oration on the Usefulness of the Mechanical Method in Medicine," *Ibid.*, p. 95.

[37] *Ibid.*

[38] *Ibid.*, pp. 96–97.

> their *Functions*; which are all performed by *mechanical Laws*, and by them only are intelligible.[39]

Returning to the oration, Boerhaave concluded that

> the human body is a mechanism, the solid parts of which are either vessels, capable of encompassing, directing, changing, separating, collecting, and secreting liquids,–or mechanical instruments which through their form, firmness, and through the way in which they are joined, are able to sustain other parts, or to produce certain movements.[40]

The causes of disease may well turn out to be mechanical in nature as well[41] and students were advised to study the works of Borelli, Bellini, and Pitcairne.[42] But then, what of the relation of chemistry to medicine? Chemical experiments give

> a very limited survey of isolated processes, insofar as they produce under some clearly defined conditions something that can be perceived by the senses . . . This science is therefore most necessary for medicine, inasmuch as it yields an abundance of observed facts and provides us with an excellent compendium of methods of observation. . . . However, these people [the iatrochemists] should not be so haughty as is their wont, and not utter the vain boast that they possess the whole treasure house of medical science through the study of chemistry alone.[43]

Thus Boerhaave admitted the value of chemistry to medicine, but he rejected the broad claims of some medical chemists that it was the very foundation of medicine.

---

[39]Herman Boerhaave, *Academical Lectures on the Theory of Physic. Being a Genuine Translation of his Institutes and Explanatory Comment* . . ., 6 vols. (London: W. Innys, 1742–1746), vol. 1, p. 81 (notes on passage continue to p. 85).

[40]"Oration on the Usefulness of the Mechanical Method in Medicine," *Boerhaave's Orations*, p. 102.

[41]*Ibid.*, p. 113.

[42]*Ibid.*, p. 110.

[43]*Ibid.*, p. 105.

Boerhaave had to wait six years for a vacancy and the promised chair. Finally, in 1709, Peter Hotton, professor of medicine and botany, died. The month following Hotton's death, Boerhaave was appointed Professor of Medicine and Botany . . . and on March 20th he gave his professorial inaugural address "On the Simplicity of Purified Medicine." Although genuinely interested in botany, this was a subject he did not know well and, as a result, the oratio refers little to this science. Instead, he referred back to themes he had discussed earlier. In presenting a short history of medicine, Boerhaave praised Hippocrates as the great founder of the subject. But what of Galen? Apart from his use of Hippocrates, "the rest is paltry, lacking in brilliance; it collapses as a mere nerveless bulk."[44] The works of the Arabs are vast storehouses of medical knowledge, but "it is only Aristotle and Galen who speak in the writings of the Arabs, albeit in a different language."[45]

Among the moderns many physicians have followed Descartes, but "they have fashioned causes through mere speculation."[46] The Cartesians "deduced most things from fictitious causes and, putting their trust in mere generalities, made an enormous leap to particulars." Because of this, "their work is so useless to the physician that medical science may safely throw off this great burden."[47]

The chemists have surely made valuable discoveries, "but as soon as they attempted to take these as a starting point for setting laws to which all bodies would be subject, they erred most shamefully."[48]

> We immediately hear a senseless din about elements! Such silly falsehoods about the ferments! Fictitious reactions of effervescing bodies! Salts are opposed to the mechanism of Nature which always performs its actions unassisted! Prophecies that breathe nothing less than immortality! Everything brought forth by chemistry is beneficial! And the work of the physician is subordinated to these fables!
>
> I maintain that we meet here with a rich source of infinite error!
>
> If all this is carefully removed from the work of Paracelsus, Van Helmont, and Tachen, from the writings of the ancient chemists,

[44]"Oration on the Simplicity of Purified Medicine," *Ibid.*, p. 130.

[45]*Ibid.*

[46]*Ibid.*, p. 131.

[47]*Ibid.*, p. 132.

[48]*Ibid.*

Boerhaave lecturing at Leiden. Courtesy of the National Library of Medicine.

> and if we carefully select from them what is useful for medical science–I ask you: is there much left?
>
> Upon my word–merely some few and simple facts which the expert Boyle had made clear to all, by means of his diligent effort, careful discretion and unblemished integrity, in plain language.[49]

Thus we see that much of what is new in medicine since Harvey may be traced to the Cartesians and the chemists and both of these schools pale before the ancient truths of Hippocrates.

In 1714 Boerhaave was appointed Rector Magnificus (or Vice-Chancellor) of the University. A year later–at the end of his term–he gave another formal address, this time "On the Achievement of Certainty in Physics." Here he centered his discussion less on medicine than on the state of science. This is a Newtonian address that is directed against the Cartesians. This is nothing new in itself, but in the past Boerhaave had spoken of the effect of the Cartesians' approach on medicine. In 1715 he attacked its influence on science in general. Above all, he rejected the assumption of first principles as a basis of knowledge. He praised Descartes for his mathematical works on geometry and dioptrics, pointing out that others building on his work "have made more progress in mathematics in the next seventy years, than this science advanced in the whole period that lies between Archimedes and Descartes."[50]

> Yet as soon as you come down from the mathemetical writings to the physical works of this eminent author, it is very difficult to believe that such diverse products could originate from the same man. For indeed, if one has the opportunity seriously to scrutinize what he put together in writing on the impenetrable infinite, at first in repose, afterwards in motion; on the laws of motion; on the origin of the world; on the power of elasticity; on the nature of the magnet; on grooved bodies and apertures that admit or reject the latter; on the origin, the structure, and the actions of the human body–then you will be astounded that such ideas should have come forth from an author who outshone everybody else in mathematical acumen! You must search for Descartes in Descartes! You will be justified in retorting against him what he himself rightly said about the Peripatetics:

---

[49] *Ibid.*, pp. 132–133.

[50] "Discourse on the Achievement of Certainty in Physics," *Ibid.*, p. 174.

> that no problem in physics can be explained by means of such principles, and that these interpretations of nature themselves show up the imperfections of assumed first principles.[51]

Huyghens seemed to be a similar case: a brilliant mathematician, but one who departed from the mathematical approach when dealing with physical problems.[52]

The true path to knowledge involves a union of mathematics and observation. For Boerhaave, Francis Bacon deserved much credit. "I shall declare ungrudgingly that for all progress achieved by natural history from the beginning of the sixteenth century until this present moment we are indebted to the warnings, the precepts, and experiments of this man; his imperishable memory will be eternally honoured by a grateful world."[53] Following Bacon we see the founding of the Royal Society "to the greatest advantage of physical science! What authors, what discoveries has it brought forth! Men like Boyle, Wren, Wallis, Hooke, Grewe, Newton and Halley."[54] And then the academies in Florence, Paris, and Berlin have all contributed to the rapid growth of a true experimental and observational science joined with mathematics.

Also in 1714 Boerhaave accepted the duties of another professorial chair, this time that of clinical medicine. And when Jacobus Le Mort (1650–1718), the professor of chemistry, died, Boerhaave took over that position as well.[55] Le Mort had managed the chemical laboratory since 1690 and had been promoted to professor in 1697, although the appointment was not made official until 1702. However, as a young lecturer Boerhaave had obtained permission in 1702 to give private lectures in chemistry and these proved to be far more popular than those of Le Mort. The two men differed considerably in their approach to the subject. Le Mort presented the subject in a manner similar to that of the many seventeenth-century textbook authors, that is, as a subject primarily of value to the physician interested in the preparation of chemical medicines. Boerhaave, on the other hand, as a Newtonian, had far less interest in the medical value of chemistry and saw it as an independent science.

---

[51] *Ibid.*, pp. 174–175.

[52] *Ibid.*, p. 175.

[53] *Ibid.*, pp. 177–178.

[54] *Ibid.*, p. 178.

[55] I have discussed Jacobus Le Mort and the introduction of chemistry at Leiden prior to Boerhaave in "Chemistry and the Universities in the Seventeenth Century," *Academiae Analecta: Klasse der Wetenschappen, 48* (1986), 15–33 (30–31).

On August 8th, 1718, Boerhaave gave another inaugural address on assuming the chair in chemistry. This time his title was "On Chemistry Purging Itself of Its Own Errors." Here he spoke of the many difficulties and errors to be found in the older literature, but he saw hope for the future in the works of recent authors. At no time did he damn all the work of his predecessors, but he surely was critical of many aspects of the earlier literature. Thus, he objected to the frequent connection of chemistry to religion that was common to these authors.

> How I wish that they had not recklessly ventured into the mysteries of official religion! Would that they had not polluted everything by mingling the vain wantonness of their unnatural speculation with these mysteries, hideously defiling them! If these madmen had only practised self-restraint and had not attempted to explain the Sacred Books themselves in terms of chemical principles elements! One feels ashamed, acknowledging that all this can be alleged of them and cannot be refuted. When we read the works of Paracelsus, Van Helmont, the Rosicrucian Brothers, and others, we are horrified and embarrassed by the silly nonsense into which the crowd of chemists plunges with sinful trespass, rashly devising all kinds of fancies. What fables, superstition, ignorance! Indeed, hardly anywhere can more raving madness be found.[56]

To prove his thesis, Boerhaave presented a brief history of chemistry. Chemical knowledge, he argued, began with the work of miners who were beset with dangers of all sorts in the subterranean world in which they lived. Uneducated and stupid, these men made up fabulous stories to explain the terrors they witnessed in the mines. These stories have continued in the literature. Those interested in the chemical art combined it with magic and fanciful writings of Plato and in the end the subject expanded its limits ever further.

> One error forges the next one! Those people regarded the Books of Moses, Solomon's writings, the Revelation of St. John as a description of the art of making gold. I pass over in silence the heinous and abominable interpretation of the Holy Trinity and of other recondite

[56]"Discourse on Chemistry Purging Itself Of Its Own Errors," *Boerhaave's Orations*, p. 195.

> religious mysteries. . . . Eventually they went to such lengths in their madness that–who will believe this?–they ended up by converting the Sacred History itself, the very miracles by means of which the Holy Authors supported the truth of their teaching, into maxims of alchemy.[57]

Still, more sober chemists, in particular Roger Bacon and Robert Boyle, sought to divorce chemistry of its far-flung claims of religious certainty.

Today, he continued, there remain many who claim that chemistry is the very foundation of medicine. The physiology of the body has been explained by them through the processes of "effervescence, opposing salts and heating sulphur, of fermentation, putrefaction, generation, transmutation, and precipitation . . ."[58] "The human body became a chemical laboratory, an arena in which the chemists could stage their games; and they introduced the clash between opposed salts, their raging strife, peace, reversal, growth, destruction and action–as if they were fully justified in doing so."[59] Let us take an example:

> The human stomach became a Hermetical crucible in which the tepid acidity of ferment causes the food to ferment in its turn. Then it becomes sour, and effervescent chyle springs forth with a fervid onset; it rushes to meet the alkaline fuel of the bile–and then you have matches of pugilists, which are stirred up by the firebrand, the spectator of the fight, to wit, the fluid from the pancreas. The work is in full swing, the captains rage against each other, on all sides attended by their satellites. Some of them rush of their own accord into the hiding-places of the milky veins, and through their own impetus they penetrate in the blood-stream, via the meanderings of the milky brook. Yet they find no quiet dwelling place here either; at once hidden enemies from their ambushes assail these newcomers and so a new battle has to be fought. Meanwhile others, spurred on by their eagerness in the pursuit of the fugitives, win their way through the narrow passages into the bed of the blood-river; they come upon their enemies and resume the fight. And in the course of this renewed contest they spread over a large space and then force their way into the first chamber of the heart's sanctuary. Raging with

[57] *Ibid.*, p. 195.

[58] *Ibid.*, p. 200.

[59] *Ibid.*, p. 206.

> the heat of the recent battle they then fly to and fro through the isthmuses of the lungs, and rush through myriads of little canals. This is not yet the end. For the paths meet again, bringing the scattered troops together, and send them forth into another secret vault of the effervescent heart. Hence they break through the obstacles with renewed vigour, and having roved along all the circuitous ways and pitfalls of the body, they return, still undaunted, to the sanctuary of the heart.[60]

How did such explanations arise? In large part, Boerhaave explained, it was due to the persuasive nature of Paracelsus, whose followers had great success in converting the physicians of Central Europe and much of France. Later van Helmont, "a man with a keen mind, learned in the liberal arts, and famed for his erudition and eloquence,"[61] rejected scholastic medicine and put chemical medicine in its place. At Leiden Franciscus de la Boë was convinced of the value of chemical medicine.

> The result was that chemical medicine became dominant throughout the whole length and Breadth of Europe. Some people still had their doubts, but the doctrine was accepted by almost everyone, as soon as Otto Tachen–a daring, magnanimous, and fortunate man–became its champion, publishing three indubitably learned and elaborate books on the subject. The idea that nature acted through chemical means and that human life was determined by them; that by these means only, and through no other cause, all motions were evoked, directed, augmented, diminished, and quieted down; and that through their variety the effects in the whole universe as well as in the human body were achieved, and that nothing came about without them–all this was drummed into every head.[62]

---

[60] *Ibid.*, pp. 207–208.

[61] *Ibid.*, p. 206.

[62] *Ibid.*, p. 207. In his medical lectures to his students, Boerhaave wrote that van Helmont saw that neither the system of Paracelsus nor the disputes of the schools would lead to truth. "But the Remedy which he applied was in effect worse than the Disease, for he founded Physic upon Chemistry, that he would no other certain Way either to its Theory or Practice. After *Helmont*, *Sylvius de la Boe* first introduced Chemistry in the University at Leyden, and persuaded the Stewards to build a publick Elaboratory for it. Chemistry is certainly a good Servant to Physic, but it makes as bad a Master." Boerhaave, *Academical Lectures*, *1*, p. 40.

Now, he continued, we see the work of more careful chemists who have shown that chemical reactions are not the same as those that occur in the human body and that it is

> wrong to draw conclusions about the cure of patients from the refining of metals. It was proved that the human organism deals with the causes of illness through means that cannot be imitated at all by the science of chemistry; that life and health depend on so many various, intricate, subtle, and wholly intangible causes that one could not really expect all the great promises held out by chemistry to be fulfilled. Even so, when the chemists had established these and innumerable other facts of the same kind by means of true experiments, the faults that had formerly been committed in the field of medicine were successfully corrected.[63]

It seemed clear then that chemists themselves had corrected these errors of the most avid iatrochemists.

In the years following his acceptance of the Chair in Chemistry, Boerhaave suffered from serious illness, first in 1722, and again five years later. Early in 1729 he thought it wise to resign from two of his four professorships. At a meeting of the Curators in February, he asked to be relieved of the chairs of botany and chemistry. With considerable reluctance his request was granted. In April he addressed the University once more. This was a more personal presentation recalling his life and the joys he had found in the pursuit of both botany and chemistry over the years. Here, in contrast to his earlier criticism of the alchemists, he took a more moderate tone.

> I at last seemed to discern with exceeding joy that the ancient Hermeticists had been most truthful philosophers. I gained the impression that Nature had never been more clearly revealed than through their endeavours, and that no other philosophers had written more plainly–provided that the scientist, being a fellow worker and having his share of the craftsmanship consequent upon such practice, understands their language; in this nothing but the results are set out of experiments that are beyond the understanding of the general public.[64]

---

[63]"On Chemistry Purging Itself of Its Own Errors," *Boerhaave's Orations*, p. 210.

[64]"Academic Discourse, Delivered by Herman Boerhaave When He Officially Resigned His Professorships in Botany and Chemistry, Having Obtained an Honourable Discharge, On 28 April 1729," *Boerhaave's Orations*, p. 233.

Indeed, he added that he had "set aside some important [chemical] questions for the leisure Of my old age, if God will grant this to me . . ."[65]

Although Boerhaave had lessened his teaching load somewhat, he remained extremely busy with his classes in medicine and clinical teaching. He never lost his interest in botany and his work in chemistry became more intense in his final years. In addition, the University appointed him Rector Magnificus a second time for the year 1730–1731. On the final day of his term (2 February 1731) he gave his final public oration, this time a "Discourse on Servitude as the Physician's Glory." Here nature is presented as the true model for the physician, who should be its servant. "It is the physician's duty carefully to observe Nature's laws, to follow them faithfully, as a servant obeys the commands of his mistress."[66] Hippocrates is offered as the best example of such a physician since

> he acknowledges Nature as the healer of diseases, and he asserts repeatedly that she is sufficient to herself in all situations; that she imparts, governs, and maintains the vital functions; that she never undertakes anything foolish or aimless, and favours the normal; that it is the duty of the physician, her servant, to observe, to recollect, to compare, and to reason solely on the basis of such data; so that he may generously place at her disposal whatever is needed; remove obstacles, with provident care, help Nature and follow her lead; and that thus, by never going against her, but by following her as a faithful guide, we are granted a long span of healthy life.[67]

We may be surprised to see Boerhaave at this point turning to van Helmont as a second example of a physician to emulate. But now he characterized him as "a great man, even if he overrated the importance of his chemical science."

> He acknowledged Nature as being the generative force that sets everything in motion by means of which she tries to get rid of the harmful substances she has absorbed, when a tiny spark has extended into a vast and destructive fire–such as the plague, pustules, ulcers, venereal diseases, poisons, and the contagion of rabies;

---

[65] *Ibid.*

[66] "Academic Discourse on Servitude as the Physician's Glory," *Boerhaave's Orations*, p. 261.

[67] *Ibid.*

> he added, on the basis of Paracelsus' theories and his own, that there are substances, minute particles, which give off exhalations that are salubrious for human life and will cure diseases otherwise irremediable.[68]

Finally he turned to Thomas Sydenham who brought forth "from his own store his Personal observation of Nature whenever they were needed–observations which had previously been collected there, laid up and put in order. And because these three men [Hippocrates, van Helmont and Sydenham] served Nature faithfully indeed, she has given them the highest award she has to offer."[69]

BOERHAAVE THE CHEMIST: THE FINAL YEARS

After relinquishing the chairs of botany and chemistry, Boerhaave continued to give his normal lectures in medicine almost to his death in 1738 and to publish in many of the fields of interest to him. However, as he had promised in his final oration, he began to spend more time on his chemical experiments. His long interest in the subject stemmed from his youth. He had given private lectures on chemistry from 1702 and he had willingly accepted the chair in chemistry at the death of Le Mort in 1718. However, he had published nothing on this subject until copies of his notes taken by students were

---

[68] *Ibid.*, p. 262.

[69] *Ibid.* It is interesting to note that Sydenham's view of the relation of chemistry to medicine is close to that of Boerhaave. In his *Treatise of the Dropsy*, Sydenham wrote that "as *Hippocrates* censures such as spend more time in studying the more curious and less useful parts of anatomy, than in practical observations, and consequently in studying nature; so there is the same reason in this age to censure those, who maintain that the art of medicine cannot be better improved, than by new discoveries in *chemistry*. For though gratitude requires us to own the helps we have received from chemistry, whence we have been supplied with some good and well-adapted medicines to answer indications, amongst which the infusion of *crocus metallorum*, the vomitive under consideration, is no contemptible one; so that for this reason chemistry is to be held a very useful art, provided it be restrained within the bounds of pharmacy: yet they are guilty of a fault, or at least a mistake, who work their imagination up to such a pitch, as at length to suppose the chief deficiency of medicine to be a want of efficacious remedies, only procurable from chemistry." *The Works of Dr. Thomas Sydenham, Newly made English from the Originals: Wherein the History of Acute and Chronic Diseases, and the Safest and most Effectual Methods of treating them, are faithfully, clearly, and accurately delivered. To which are added, Explanatory and Practical Notes, From the best medicinal Writers*, The Second Edition, Revised, Corrected, and Enlarged with several Additional Notes by John Swan, M.D. (London: Edward Cave, 1749), p. 473

printed, to his dismay. In 1724 the *Institutiones et experimenta chemiae* appeared with a Paris imprint. This work was then reprinted in Amsterdam and Venice; in 1727 Peter Shaw and Ephraim Chambers provided an English translation. Boerhaave had authorized none of these books and he felt that they all contained errors. They angered Boerhaave because he saw his students following his lectures with the volumes open on their laps. Accordingly he began to adapt his notes to publishable form. His version appeared first in 1731 in Leiden and it was then republished the following year in Tübingen, Basel, and again at Leiden. An English translation by Timothy Dallowe appeared in 1735. Many editions followed and there is little doubt that Boerhaave's text became the standard work on chemistry in the middle decades of the eighteenth century.[70]

Although his chemical lectures attained great popularity, the text itself in many respects reflects the traditional organization of such printed guides to chemistry printed over the preceding century. It begins with a history of the subject and then goes on to chemical theory, discussing minerals, vegetables, and animal substances. From these he turns to laboratory operations and the properties of Fire, Air, Water, and Earth. The second part proceeds to take up chemical practice, that is, how to prepare chemicals in the laboratory.

Boerhaave's chemical discoveries cannot be compared with those of Boyle, but Lindeboom mentions his isolation of urea from urine, the production of water by burning alcohol, and his regular use of Fahrenheit's thermometer.[71] However, the best summary of Boerhaave's chemical text appears in Partington's *A History of Chemistry.* Here we find an account of Boerhaave's attempt to determine the possible weight of fire by heating iron and copper bars, the types of salts and their solubilities in water, his views on atomism, on alkalis and acids, and a large number of other observations placed in a rather traditional setting.[72] The book is clearly an experimental and observational text, and because of its comprehensive nature, its popularity in the first half of the eighteenth century is easily understandable.

---

[70]For a discussion of the early editions of Boerhaave's chemical text, see Lindeboom, *Herman Boerhaave*, pp. 175–184. To ensure the authenticity of the *Elementa chemiae*, Boerhaave signed every copy of the first edition, *Ibid.*, p. 182. J. R. R. Christie has compared the historical section of the English translations of *The New Method* by Peter Shaw and Timothy Dallowe in "Historiography of Chemistry in the Eighteenth Century: Hermann Boerhaave and William Cullen," *Ambix*, *41* (1994), 4–19.

[71]*Ibid.*, pp. 331–340.

[72]J. R. Partington, *A History of Chemistry*, vol. 2 (London: Macmillan & Co./New York: St. Martin's Press, 1962), pp. 747–759.

Boerhaave's academic orations have already indicated his attitude toward chemistry and have given some understanding of his earlier views. For him it was a subject of considerable importance on its own because "chemistry teaches us the changes which bodies suffer of themselves, and when applied to Fire; for Experiments themselves teach us no Falacy . . ."[73] In 1718 he had pointed out that earlier errors in chemical texts had been corrected by recent chemists. But who were these erroneous authors? Paracelsus may have been a successful surgeon who knew well the practice of medicine, including the uses of metals. However, "these take in his whole merit: and were the matter of all his glory: the rest was empty smoke, and idle ostentation."[74]

And yet, Paracelsus and van Helmont ranked among Boerhaave's favorite authors. His first biographer, William Burton, affirmed that Boerhaave "had read over carefully PARACELSUS four, and HELMONTIUS seven times: the latter was his favorite . . . but he rather esteemed him as a philosopher than a physician, having found the efficacy of his most applauded remedies to fall so very short of their solemnly avowed success." But even though van Helmont was to be blamed for attempting to make medicine subservient to chemistry, Boerhaave knew his work well enough to be able to "repeat some whole chapters of van Helmont almost verbatim."[75] Referring to van Helmont in the *New Method of Chemistry*, Boerhaave stated that a "man of his genius and views, seeing so many more powerful remedies produced by chemistry; naturally enough began to imagine, that all might be true, *Paracelsus* had advanced; and thus took the contrary byass, run counter in every thing to the *Galenic* school, banish'd every thing of that kind out of medicine, and reduced the whole art to principles of chemistry."[76] It is clear that although he had the highest respect for van Helmont even late in life, Boerhaave continued to reject the concept of iatrochemistry. There were, indeed, many authors–Francis Bacon and Robert Boyle ranked high among them–who had shown the value of chemistry for medicine, but "I do not rank *Francis de la Boe Sylvius*, and *Otto Tachenius*, with them, by reason their too

---

[73]Boerhaave, *Academical Lectures*, *1*, p. 44.

[74]Hermann Boerhaave, *A New Method of Chemistry; Including the Theory and Practice of that Art: Laid down on Mechanical Principles, and accomodated to the Uses of Life. The Whole making A Clear and Rational System of Chemical Philosophy. To which is prefixed A Critical History of Chemistry and Chemists, From the Origin of the Art to the present Time*, translated by P. Shaw and E. Chambers (London: J. Osborn and T. Longman, 1727), *1*, p. 28.

[75]William Burton, *An Account of the Life and Writings of Herman Boerhaave . . . In Two Parts, with an Appendix* (London: Henry Lintot, 1743), pp. 157–158.

[76]Boerhaave, *New Method* (1727), *1*, p. 32.

great fondness and attachment to chemistry, led them to consider it rather than the servant of physic."[77]

Chemistry was clearly not to become the theoretical basis of medicine, but there is little doubt that Boerhaave thought that chemistry was a useful subject for the physician. Discussing this point in his lectures, he stated that chemistry brings light to physiology. Mechanics will not help in a study of the matter of the body, but chemistry will help through analysis. Pathology is inexplicable by any other means than chemistry and it is useful as well in semeiotics "or *the doctrine of signs*" which "is principally concern'd about the pulse and urine." "Hygiene . . . instructs us in *the way of preserving health by diet:* and this entirely depends on chemistry" because it can tell us which foods are best for us. Chemistry shows us "that the natural heat of our bodies will turn vegetables acid, and putrefy animal substances. Whence we may learn, that in case of an alkaline disposition of the blood and juices, when the urine appears red, smells ill, and tastes alkalious; vegetable acids are the proper diet, and animal food prejudicial." These and other observations can only be determined through chemistry. Finally he turned to therapeutics and wrote that the cure of disease is also dependent on chemistry.

> When a patient is seized with a burning fever, chemistry immediately informs us, that, from the additional heat, the salts of the blood become sharper; and as the heat increases, are render'd alkaline, and the oils more volatile, and exhalable. Whence we infer, that acids ought to be prescribed, to prevent the putrefaction that is bringing upon the juices: and this indication could be fetched from nothing but chemistry.[78]

There are many cases where the new chemically prepared medicines have proved to be valuable in the cure of specific diseases and because of this we can no longer rely exclusively on the ancient medical texts. "As we are acquainted with a great variety of things unknown to the antients, the arguments drawn from their physic, without the late assistances, and particularly that of chemistry, are not to be relied on."

[77]Hermann Boerhaave, *A New Method of Chemistry; Including the History, Theory, and Practice of the Art, Translated from the Original Latin of Dr. Boerhaave's Elementa Chemiae, As Published by Himself. To which are added Notes; and an Appendix, Shewing The Necessity and Utility of Enlarging the Bounds of Chemistry. With Sculptures.* By Peter Shaw, 2nd edition, 2 vols. (London: T. Longman, 1741), *1*, p. 177.

[78]Boerhaave, *New Method* (1727), *1*, p. 197.

But what of the alchemists? In his academic oration of 1729, Boerhaave referred to the Hermeticists as "most truthful philosophers" who had "written plainly," withholding only some information that was unfit for the general public.[79] He had read widely in this literature and had a high regard for the authors, especially Geber. When he read understandable texts he had found them to be correct, so why should he doubt them when he did not understand them? After all, these adepts "have shown them-selves so much better skill'd in the art than myself."[80] "Thus much I have long ago had a mind to say, concerning the knowledge of the true alchemists in physics; lest such skilful artists should be condemn'd by incompetent judges."[81] True, the alchemists made some grievous errors, but they sought an understanding of nature through experiment, as do all true scholars.

Boerhaave did seek a more complete understanding of the alchemical texts. He considered these works to be founded on experiments and he felt that it was necessary to decipher the language in order to uncover the chemical truths contained within. In 1729 he had stated that although he was resigning from his professorship in chemistry, he had "set aside some important chemical questions for the leisure of his old age."[82] These experiments were to be related to alchemy. In the 1733–1734 and the 1735–1736 issues of the *Philosophical Transactions of the Royal Society of London*, there appeared three articles by Boerhaave on his mercurial experiments.[83] In sending the first of these to his friend, Cromwell Mortimer, in London, Boerhaave wrote that "It is a source of annoyance to me that the ignorant reject the old writers, while I in fact prove that the latter have contended with brevity and lucidity those very things which I, by means of years of experimenting, have learnt to be

[79]*Boerhaave's Orations*, p. 233.

[80]Boerhaave, *New Method* (1741), *2*, p. 202.

[81]*Ibid.*, p. 203.

[82]*Boerhaave's Orations*, p. 233.

[83]H. Boerhaave, "De mercurio experimenta," *Philosophical Transactions of the Royal Society*, *38* (1733–1734), 145–167; *39* (1735–1736), 343–359, 368–376. The first of these appeared in English as *Some Experiments Concerning Mercury . . . Translated from the* Latin, *Communicated by the Author . . . to the ROYAL SOCIETY* (London: J. Roberts, 1734). Abridged translations will be found in *The Philosophical Transactions of the Royal Society of London From Their Commencement, in 1665, to the Year 1800; Abridged, with Notes and Biographic Illustrations by Charles Hutton, George Shaw, and Richard Pearson*, 18 vols. (London: C. and R. Baldwin, 1809), *7*, pp. 619–623; *8*, pp. 93–95, 98. A French version of the first article will be found in the *Memoires . . . de la Academie Royale des Sciences*, 1734 volume (Amsterdam: Pierre Mortier, 1738), pp. 739–758.

the truth."[84] And, a year later, writing to his good friend, Joannes Baptista Bassand, Archiater to the Emperor, he wrote that

> I am liberating my old age, the disabilities of which are worsening, from the academic and domestic activities that you yourself formerly observed. At home I devote both day and night to chemistry. If you wish a specimen of my foolish zeal, see the enclosed work on mercury that I have given to the English, in order to defend Geber's opinion. I have offered another work with which equal pains have been taken, to the French, and the experiments with gold and silver will be published shortly: these confirm the real truthfulness of the old Alchemists and reveal the instability of the doctrine of many men of more recent times.[85]

In his first communication on mercury to the Royal Society, Boerhaave opens by lauding the important discoveries due to chemistry, adding that "the most ancient Alchemists far surpassed the rest in their Accounts of the Nature of Things." Geber and the writers closest to him wrote the most plainly and "no other Men whatever have so strictly and obstinately labour'd in the Search after natural Things, or have taken such great Pains to turn Matter, thro' all the various Modes of Enquiry, as the Alchemists."[86] But the same alchemists have been accused of hiding their knowledge "in Obscurity and Darkness." Indeed, many say that what they describe is impossible. The alchemical authors state that their work is based on experiments, but that their mysteries are not for the profane persons. For this reason it is essential that these secrets should be hidden from those not initiated in the art.[87]

On the basis of his reading of this literature, Boerhaave concluded that the alchemists agreed that (a) metals are generated and grow like other natural things, (b) there are metallic seeds with seminal powers, and (c) that these seeds need a proper matrix and a suitable heat to mature to a metal in a definite period of gestation.[88] This matrix is "thick, hard, impenetrable, without Mixture, closed up on all Parts, and resembles Glass."[89] The process

---

[84]G. A. Lindeboom, editor, *Boerhaave's Correspondence*, 3 vols. (Leiden: E. J. Brill, 1962, 1964, 1979), *1*, pp. 210, 211. Boerhaave to C. Mortimer, 8 April 1734.

[85]*Ibid.*, *2*, pp. 334–335. Boerhaave to J. B. Bassand, 20 July 1735.

[86]Boerhaave, *Some Experiments*, p. 6.

[87]*Ibid.*, pp. 7–8.

[88]*Ibid.*, pp. 9–10.

[89]*Ibid.*, p. 11.

itself is similar to growth in the animal and vegetable kingdoms where foods nourish the body. So too, "the vivificating Seed of growing Gold, having got a proper Food, in a fit Matrix, by the Help of a suitable and convenient Heat, digests the same into its own particular Nature."[90] Gold then is produced out of a matter different from itself.

Again turning to the alchemical authorities, Boerhaave stated that most agreed that mercury "is the common matter of all Metals." Specific metals arise from the addition of the seed to the quicksilver. Thus, "every Metal, when the Quicksilver and this Metallific Power (which they call Sulphur) are maturely, and, as it were, thoroughly boil'd and concocted, is brought to a perfect Species of each such Metal. And from hence, That every Metal is again resolved into these two Principles, Mercury and Sulphur."[91] If pure mercury could be obtained there would be no problem producing any metal. But there is a flaw in common quicksilver that is very difficult to remove. This, then, is Boerhaave's task: "I have endeavour'd to learn by Experience, by what Artifice a pure unmix'd Mercury might be obtained."[92]

In his search for a purified mercury, Boerhaave discussed numerous experiments. He heated mercury for more than fifteen years, but found no change in the metal. In another case he distilled and redistilled mercury more than five hundred times. At best he found small amounts of black and red powder that reverted to the metallic form on trituration. In his eighth experiment he took seven drachms and thirty-seven grains of mercury powder, which he distilled, and found seven drachms of the "purest mercury" in the receiver, while the retort had fifteen grains of dark, fixed powder. His conclusion in this case was that mercury was changed into powder by fire alone. He explained this in alchemical terminology: "Thus the Serpent that has bitten itself dies" and "It arises again more glorious from death."[93] He became convinced that mercury persists in the fire and that while it may acquire different degrees of fixedness, it never reaches the degree of fixedness of gold or silver.[94] "Therefore it does not appear, by these Experiments, that from Mercury and Fire so conspiring, any known Metal is produced" and further, "Fire, by these Experiments, is not demonstrated to be the Sulphur of the

[90] *Ibid.*, p. 10.

[91] *Ibid.*, p. 13.

[92] *Ibid.*, p. 14.

[93] *Ibid.*, pp. 38–40.

[94] *Ibid.*, pp. 43–44.

Philosophers that fixes Mercury into Metals."[95] But there may be another approach. Geber had written that pure mercury is heavier than gold. Would it be possible then to thicken mercury "into the weight of Gold? Would it then be live Gold, or the Mercury of the Philosophers?"[96]

Following the work of Franciscus Mercurius van Helmont, Johann Becher, and Isaac Hollandus, Boerhaave next sought to extract mercury from other metals. In his *Paradoxal Discourses*, the younger van Helmont stated that lead dissolved by alkalies, oil, and salt becomes changed into a "volatile running Mercury, which can no more endure the fire, as before, but is cold and running like water, and without a metalline form."[97] Boerhaave was unable to obtain a mercury by this process. Nor was he able to obtain a purified mercury by treating amalgams of mercury and gold even though he distilled them 877 times.[98] Although these experiments did not result in the desired conclusion, it seems likely that Boerhaave had not given up his hope that the search for a purified mercury would be successful.

Nor need we be surprised to find Boerhaave's deep interest in the *Alkahest*, the famed universal solvent described by Paracelsus and the elder van Helmont. A discussion of the references to it in the works of these two authors formed a part of his lectures on chemistry and he admitted to making over one hundred experiments during two years' time in his search for this menstruum.[99] Here too he had to admit defeat and made the admission that since the times of Paracelsus and van Helmont this search "has employed the studies and labours of so many famous chemists, with great expences, to discover."[100]

Boerhaave's interest in the more arcane aspects of chemistry may account for the fact that he became associated with the famous wound powder that supposedly cured by anointing the weapon rather than the wound it made. Again, this is a subject frequently discussed in the seventeenth

---

[95] *Ibid.*, p. 45.

[96] *Ibid.*, pp. 51, 54.

[97] F.M. Van Helmont, *The Paradoxal Discourses of F.M. Van Helmont, Concerning the Macrocosm and Microcosm, or the Greater and Lesser World And their Union. Set down in Writing by J.B. and now Published* (London: J. C. and Freeman Collins for Robert Kettelwel, 1685), Part I, p. 111.

[98] See the discussion of Boerhaave's second and third papers on the experiments on mercury in the abridged *Philosophical Transactions* (note 82 above), *8*, pp. 92–93.

[99] Boerhaave, *New Method* (1727), Part I, pp. 362–372 (371).

[100] *Ibid.*, p. 372.

century.[101] In 1743 there appeared *An Essay on the Virtue and Efficient Cause of Magnetical Cures. To which is added, A new Method for curing Wounds without Pains, and without the Application of Remedies. Hitherto kept as a Secret in private Families. Written Originally in Latin, by the Famous Dr. Herman Boerhaave, Professor at the University of Leyden in Holland, But never Published till now.* This book appeared in London without naming the publisher and it has been questioned as spurious. However, whether or not this is true, I believe that there is enough material on Boerhaave's alchemical interests to question such an assertion. The author suggests that cures by the sympathetic powder may be a mystery to some, but it is "nothing else, but a Science of the Sympathies and Antipathies between superiour and inferiour Things."[102] The loadstone and the effect of the sun on flowers is presented as evidence and these are said to be similar to the interaction of the macrocosm and the microcosm.[103] We must seek for the signatures that indicate the heavenly influence on earth and teach us "the true magnetical Virtue of the Firmament, and the means to resist Diseases caused by such Impressions."[104] We must be careful not to call into question all cures of this sort and even if the cause of such a cure is unknown there is no need to ascribe it to the devil. In fact all parts of the world are joined together by the universal spirit that bestows virtue on everything.[105] This being true, "who can doubt of the Possibility of curing Wounds without applying any Remedy, only by anointing the Instrument wherewith such a Wound is made?"[106]

Boerhaave was surely a complex individual. Although his *Elementa Chemiae* became a standard text in the middle decades of the eighteenth century, the author's veneration for Boyle, Newton, and the mechanical philos-

---

[101]The major texts are collected in the *Theatrum Sympatheticum auctum, exhibens varios authores de pulvere sympathetico* . . . (Nuremberg: Joh. Andreas Endter, 1662). I have discussed Robert Fludd's debates on this sympathetic powder in *The Chemical Philosophy: Paracelsian Science and Medicine in the Sixteenth and Seventeenth Centuries* (2 vols., New York: Science History Publications, 1977), *1*, pp. 279–290.

[102]Boerhaave, *An Essay on the Virtue and Efficient Cause of Magnetical Cures* . . . (London: n.p., 1743), p. 9.

[103]*Ibid.*, pp. 10–11.

[104]*Ibid.*, p. 22.

[105]*Ibid.*, pp. 25, 28.

[106]*Ibid.*, pp. 49–50. The procedure given for this cure is as follows: "The Instrument with which the Wound is made, or in want thereof, a piece of Wood or a little Bough of Willow steeped in the bloody Opening, is to be plunged in the Ointment and therein to be left: The wounded person, meanwhile, hath only to wash the Wound daily once with his Urine or fresh Water, and to cover it with a piece of clean Linen, always wiping away the Filth and the Wound will heal and consolidate without any Pains." *Ibid.*, p. 55.

ophy extended to van Helmont and the traditional alchemists. Indeed, his most detailed chemical experiments surely sought to establish fundamental aspects of alchemical doctrine.

GEORG ERNST STAHL (1659–1734) AND MEDICAL VITALISM

Stahl is similar to Boerhaave in that his principal work was in medicine. However, like Boerhaave he became one of the fundamental eighteenth-century authorities in chemistry. Another similarity is to be seen in the fact that both were convinced that chemistry was not to be considered an important factor in the study of human physiology. And yet, they differed profoundly in other areas. Boerhaave was a mechanist, whereas Stahl was firmly committed to the anima as a basis for life and its processes.

Today there is an increasing interest in the work of Stahl both by physicians interested in the history of medicine and by chemists seeking a deeper understanding of the phlogiston theory advocated by Stahl. Too often the views of the medical historians and the historians of chemistry find little common ground. Here our interests will center on Stahl's views of the relationship of chemistry to medicine.

Born in Anspach, Bavaria in 1659, Stahl was the son of a Protestant pastor.[107] His deep religious background remained with him throughout his life. In 1684 he received his M.D. degree at Jena, where he gave a series of lectures on chemistry that were edited by a student and published for the first time nearly forty years later as the *Fundamenta Chymiae Dogmaticae et Experimentalis* (1723). In 1687 he became a court physician to the Duke of

[107]Although the importance of Stahl's work in medicine and chemistry has long been acknowledged, there are few works that discuss the man and his contribution as a whole. Perhaps the best short treatment is that of Lester S. King in the *Dictionary of Scientific Biography, 12* (New York: Scribner's, 1975), pp. 599–606. King discussed Stahl's work in several of his other books and articles which will be referred to below. Walter Pagel's comparison of "Helmont, Leibniz, Stahl," *Archiv fur Geschichte der Medizin, 24* (1931), 19–59 is a basic text, while the nineteenth-century *Le vitalisme et l'animisme de Stahl* (Paris: Germer Baillière, 1864) by Albert Lemoine remains important. As always, J. R. Partington's *History of Chemistry* is a source that goes beyond the technical history of chemistry itself and is a basic source for bibliography and the life history of the man. Here see volume 2 (London: Macmillan, 1961), pp. 652–690. And also, concentrating primarily on chemistry, there is Hélène Metzger's *Newton, Stahl, Boerhaave et la doctrine chimique* (Paris: Félix Alcan, 1930). Currently there is a new interest in Stahl and the reader is directed to the dissertation of Ku-ming Chang (University of Chicago).

Portrait of Georg Ernst Stahl (1659–1734). Courtesy of the National Library of Medicine.

Saxe-Weimar and seven years later he accepted a position as the second professor of medicine at the newly founded (1694) University of Halle at the urging of his friend, Friedrich Hoffmann (1660–1742), with whom he was later to quarrel. Nevertheless, Stahl remained at Halle until 1716, when he moved to Berlin to become body-physician to the King of Prussia, Frederick I. He remained there until his death in 1734.

THE ATTACK ON THE MEDICAL SCHOOLS

In his search for a medicine dominated by an inherent life force or anima, Stahl rejected the traditional medical philosophies. He had little faith in the vulgar logic of the ancients taught in the universities.[108] This ancient philosophy might better be called philology, a love of words and dispute.[109] Similarly he rejected the ancient atomism of Democritus and Epicurus, while arguing that the modern mechanists were no better.[110] In his *Disquisitio de Mechanismi et Organismi Diversitate* (1706), he complained that the new physicians who sought to employ mechanical explanations in their work presented their views in vague terms that could not be made the basis of medicine.[111]

In various places he criticized Borelli, Pitcairne, and Willis. And as for Descartes, Stahl found little of value because he had considered man to be little more than an automaton.[112] These people, he wrote, speak of mechanism,

[108]G. E. Stahl, *Disqvisitio de Mechanismi et Organismi Diversitate* . . . (Halle: Sumptibus & Impensis Orphanotrophei, 1706), p. 5. In translation see *Recherches sur la difference qui Existe entre le Mécanisme et l'Organisme: Dissertation Médico-Philosophique* in *Oeuvres Médico-Philosophique et Pratiques de G.-E. Stahl*, trans. and commentaries by T. Blondin, 2nd edition (Paris/Montpellier/Strasbourg: J. B. Ballière et Fils/Pitrat/Treuttel et Wurtz, 1864), *2*, p. 122. Volumes two through six (of a planned ten) of the Blondin translations were published. The translation itself is not always literal, but it does preserve the thought of the original.

[109]G. E. Stahl, *De la Philosophie d'Hippocrate* (Halle, 1704) in *Oeuvres*, *2*, p. 185.

[110]Stahl, *Disqvisitio*, pp. 7–9. *Recherches* in *Oeuvres*, *2*, pp. 189–190.

[111]Stahl, *Paraenesis, ad Aliena a Medica Doctrina Arcendvm; Quoniam non Solvm Medicam Doctrinam Morentvr, Tvrbent, Fallant, sed etiam Aliena ab ipsa Veritate et vsv Practico, Animis Obiiciant* (Halle: Orphanotrophei, 1706), p. 14. Stahl, *De la Nécessité d'Éloigner de la Doctrine Médicale Tout ce qui lui est Étranger* (Halle, 1706) in *Oeuvres*, *2*, p. 150.

[112]Stahl, *Disquisitio*, p. 32. *Recherches* in *Oeuvres*, *2*, p. 223 (On Descartes on man as an automaton). Stahl rejected Pitcairne's views on the blood flow in the *Vera Diversitate Corporis Mixti et Vivi, et Vtrivsque Peculiarum Conditionvm atqve Proprietatvm Necessaria Discretione Demonstratio* (Halle, Magdeburg: Sumptibus Litterisqve Orphanotrophei, 1707), p. 42; *Veritable Distinction a établir entre le Mixte et le Vivant du Corps Humain* (Halle, 1707) in *Oeuvres*, *2*, p. 315. He opposed Borelli also on the heating of blood, while Willis on the same subject is called "fantastic." Stahl, *Réclamations, Défense et Indications Justificatives* in *Oeuvres*, *2*, p. 425.

mechanics, machines, and of mechanical powers, but "we are curious to know where one can find the true meaning of these terms." It would seem that today every body is a machine that functions by itself and for itself with nothing organic in its makeup.[113]

And what of the anatomists? They seek to identify ever more minute parts of the body. We do not doubt that the intimate structure of the muscles will tell us the number and arrangement of the fibers, but what value is that for a physician who must deal with a wound?[114]

As for chemistry, that science until now has been useless for true mechanical theory. Chemists base their explanations on the coagulation of the humoral mass, its acidity, the volatility of alkaline principles and on sulphuro-alkaline and fermentative properties. Indeed, the chemical physicians posit an infinity of acids in the human body, substances that it is impossible to identify.[115] Stahl clearly rejected the acid-alkali theory of his day. In short, whether one turned to the ancients, to anatomy, mechanics, or chemistry, he would fail in those attempts to explain life processes because they were all based on material considerations and had resulted in failure.

What then, is the basis of true medicine? The answer is that true medicine is properly founded on nature and requires an understanding of the *anima* or soul, which is the true principle of life. Stahl insisted on a sharp distinction between living and non-living substances.[116] A living being has motion, with an agent that regulates it. The resultant mechanism is subordinate to the goal of the organism. The governing agent is a single internal *anima* that bears some similarity to the archei of the various organs proposed by van Helmont.

For Stahl there was a need to turn from material explanations to nature itself, "the author and sustainer of our life . . . that is to say, the soul."[117] It is this immaterial *anima* that directs and accomplishes the functions of the body and its organs to maintain life. It is by motion that nature guides and oversees

[113]Stahl, *Disqvisitio*, pp. 13–14; *Recherches* in *Oeuvres*, *2*, pp. 197–198.

[114]Stahl, *Paraenesis*, pp. 16–18; *De la Nécessité* in *Oeuvres*, *2*, pp. 153–157.

[115]Stahl, *Paraenesis*, p. 22; *De la Nécessité* in *Oeuvres*, pp. 157–160. "Iatrochemistry has only been a decoy up to this day to easy and credulous imaginations." *Ibid.*, p. 164. Hélenè Metzger discussed Stahl's attack on current acid-alkali theory in *op. cit.*, pp. 113–114. Here see Stahl, *De la Nécessité* in *Oeuvres*, *2*, pp. 158–159.

[116]Stahl, *Paraenesis*, p. 12; *De la Nécessité* in *Oeuvres*, *2*, p. 147.

[117]Stahl, *Paraenesis*, p. 25; *De la Nécessité* in *Oeuvres*, *2*, p. 168.

the interests of the body. The mechanists may insist that circulation occurs through mechanic impulsion, but Stahl replied that the soul exercised its influence through circulation according to the needs of the organism.[118]

THE REACTION TO STAHLIAN ANIMISM: HOFFMANN AND LEIBNIZ

Although medical animism was to develop at Montpellier during the course of the eighteenth century, it ran into opposition at many other medical centers. Two important authors who questioned Stahl on this point were his colleague at Halle, Friedrich Hoffmann (1660–1742) and Gottfried Wilhelm Leibniz (1646–1716). Hoffmann, Leibniz, and Stahl all had a great interest in chemistry, which makes their views of special interest.

Hoffmann was educated at Erfurt and Jena, where he took his M.D. in 1681.[119] He had an early interest in chemistry and gave a series of lectures on the subject at Jena, but his *Chymia Rationalis et Experimentalis* did not appear until after his death (1748). In 1694 Hoffmann was appointed to the first chair of medicine at Halle and he chose his Jena colleague, Stahl, as a second appointment. Although Hoffmann spent three years as physician to the King of Prussia, he taught at Halle for the rest of his long life.

Hoffmann was greatly influenced by the mechanical philosophy. In his early *Fundamenta Medicinae* (1695), he insisted that medicine must be based on observation and sound reason. Indeed, nature is basically mechanical and operates through the laws of motion. "Like all of nature, medicine must be mechanical. The first principles of mechanics are matter and motion, therefore the elements thought up by the peripatetics and the chemists are mere

[118]Stahl, *Paraenesis*, pp. 25–29; *De la Nécessité* in *Oeuvres*, *2*, pp. 168–176.

[119]The account of Hoffmann in the *Dictionary of Scientific Biography* is by Guenter B. Risse, vol. 6, pp. 458–461. His chemical work is best approached through the chapter on Hoffmann and his followers in Partington, *History of Chemistry*, *2*, pp. 691–735. The works of Hoffmann were collected as the *Opera omnia physico-medica*, 6 vols. (Geneva: de Tournes, 1748; first edition, 1740). And the basic comparison of Stahl and Hoffmann will be found in Lester S. King, "Stahl and Hoffmann: A Study in Eighteenth Century Animism," *Journal of the History of Medicine and Allied Sciences*, *19* (1964), 118–130. See also King, *The Philosophy of Medicine: The Early Eighteenth Century* (Cambridge and London: Harvard University Press, 1978), pp. 118–120.

Portrait of Friedrich Hoffmann (1660–1742). Courtesy of the National Library of Medicine.

imaginings."[120] A Cartesian influence may be seen in his assertion that "our body is like a machine or automaton whose organs, varying in shape and size, are disposed and constructed in a particular order and position . . . [Indeed] life is achieved by causes that are wholly mechanical."[121]

In a short discussion of the various medical sects, Hoffmann turned to Paracelsus, who had repudiated the Aristotelian elements and had introduced the three chemical principles into medicine.[122] The latter are produced more by the fire than by nature, and, in any case, they shed no light on medicine and do not aid in the cure of disease. "We will say nothing about the absurd superstitions and arcana" he presented. As for van Helmont, he destroyed the teachings of Galen and postulated a vital principle or archeus in living beings. It is evident in the *Observationvm Physico-Chymicorvm Selectiorvm* (1722) that Hoffmann was well acquainted with the works of the iatrochemists. There are frequent references to Paracelsus, Isaac Hollandus, Basil Valentine, Glauber, Becher, and Kunckel as well as Robert Boyle, who is praised for his accurate observations and experiments. And Hoffmann, a chemist himself, gave credit to these authors who had discovered many remedies unknown to the ancients, especially the metallic preparations.[123]

As with Stahl, Hoffmann sought to separate chemistry and medicine and it is for this reason that he rejected the chemical physiology of Sylvius.[124] However, this did not prevent him from using the inorganic remedies of the chemists along with the more traditional Galenicals. He prepared an edition of Pierre Potier's (d. c. 1640) *Opera omnia practica et chymica* (1622) in 1698 that included his *Pharmacopeae spagyricae.*[125] A large collection of chemical preparations (210 pages in the edition of 1741), this work is best known for its description of the Bologna Phosphorus,[126] but it is significant also for the

[120]Friedrich Hoffmann, *Fundamenta medicinae*, trans. and introduced by Lester S. King (London: Macdonald and New York: American Elsevier, 1971), p. 6.

[121]*Ibid.*, p. 11.

[122]Hoffmann, *Prolegomena Natura, Fundamentis, et Principio* in *Opera omnia, 1*, pp. 16–17.

[123]Hoffmann, *Observationvm Physico-Chymicarvm Selectiorvm Libri III. In qvibvs mvlta cvriosvm experimenta et lectissimae virtvtis medicamenta exhibentvr, ad solidam et rationalem chymiam stabiliendam praemissi* (Halle: Prostat in Officina Libraria Rengariana, 1722), pp.):(4^{v},):():(3^{v}-4^{v},):():():(3^{r}.

[124]Partington, *op. cit.*, *2*, p. 692.

[125]I have used the edition of 1741. Petrus Poterius, *Opera Omnia Practica et Chymica, Cum Annotationibus & Additamentis Vtilissimis Pariter ac Curiosis Friderici Hoffmanni* (Venice: Ex Typographia Balleoniana, 1741). Here Potier's *Pharmacopoeae Spagyriae* with Hoffmann's annotations will be found on pp. 283–495.

[126]*Ibid.*, pp. 433–436. See the discussion by Partington, *op. cit.*, *2*, pp. 334–340.

inclusion of a short text (1666, dated 1665) on the elements and principles by Pierre Guisson, M.D. of Avignon, in which he described the views of the ancients, Paracelsus, van Helmont, and Robert Boyle.[127]

A comparison of Hoffmann's views with those of Stahl, the *Commentarius de differentia inter Friderici Hoffmanni Medico-Mechanicam: & Georgii Ernesti Stahlii Medico-Organicam* (1746) appeared posthumously. It is a work that has been examined in detail by Lester S. King, who has shown that the two authors cannot be categorized simply by the terms mechanist and animist. Here Stahl is presented as believing that matter is passive, its motion determined by the *anima*, while Hoffmann presents his belief that matter is not passive, but rather that it is provided with a motive force.[128] Hoffmann surely accepted the fact that mechanical action could explain numerous biological actions, but not all. Some, such as generation and mental functions were to be explained in terms of very subtle spirits or *virtutes* referrable to God. These operated through motion.[129] King also discussed the differences between Stahl's *anima* and Hoffmann's *virtutes.* Hoffmann did not concern himself with the immaterial, whereas Stahl did. And for the former, motion derived directly from God, while the latter argued that motion derived from the *anima* that originated from God. Hoffmann asserted that mechanical principles determined the goals pursued by living things, something denied by Stahl. Stahl also insisted that the *anima* was the principle responsible for organization in the body. For Hoffmann prime importance was given to the laws of motion expressed in the movement of particles rather than their organization.[130] In the *De differentia,* little attention was placed on chemistry, although as we have noted, Hoffmann was deeply interested in the use of chemical preparations in medicine.

Stahl's debate with Leibniz went beyond the role of the *anima* to touch on the role of chemistry in medicine. We need not discuss the latter's well-known career in science, his philosophical works, his role in the founding of the Berlin Academy, or his role in the discovery of the calculus. Here we are interested in his reaction to Stahl's rejection of chemistry in medicine and this may be related to his lifelong interest in alchemy and chemistry.[131]

---

[127]Pierre Guisson, M.D., *Epistolica Dissertatio de Anonymo Libello (Circa Abbraeviatum Verae Medicinae Genus) ubi potissimum eventilatur Principiorum Chymicorum Hypothesis* in Potier, *Opera Omnia* (1741), pp. 571–580.

[128]King, "Stahl and Hoffmann," 124.

[129]*Ibid.*, 125.

[130]*Ibid.*, 126–127.

[131]Partington, *op. cit.*, *2*, p. 485.

Leibniz was disturbed by Stahl's belief that the human soul had the power to place in motion the body to which it is united. He considered this hostile both to natural theology and to revealed religion.[132] Indeed, it was contrary to religious faith.[133] The views of both Stahl and Leibniz were published in the *Negotiosum Otiosum* (1720). Here Leibniz pointed out that both Paracelsus and van Helmont had accepted an *archeus* in man that governed appetite, irritation, and energy and that this resulted in the actions of the body. This concept had been revived by Stahl and applied to practical medicine incorrectly.

Of more interest to us is the fact that Leibniz questioned Stahl's views on the relationship of anatomy and chemistry to medicine. Noting the latter's dismissal of ever more detailed anatomical investigations, Leibniz stated that "while one ought not to ask every physician to examine all anatomical minutiae, yet I should think that it is to the public benefit that there should be no lack of those who will consider that some particular finding is of the greatest importance for them. In surgery, surely there is very great profit in an ever more refined anatomy."[134] This research, he felt, would lead in time to many cures that until now were beyond hope.[135]

As for chemistry, Leibniz objected to the fact that Stahl believed chemistry to be further removed from true medicine than anatomy. Stahl, in turn, denied that he ignored chemistry, a subject that could be praised in its own right.[136] Stahl argued that chemistry had not yet acquired a true dogmatic theory among physicians or in any systematic treatise. To date it had resulted in nothing that could be termed a true positive chemistry. It has made little progress and because of this it would be absurd to present it as the key to the other sciences.[137] In short, "I say that chemistry has little utility for the

---

[132]Georg Ernst Stahl, *Negotiosum Otiosum, seu σκια μαχια adversus positiones aliquas fundamentales theoriae verae medicae* (Halle: Litteris et Impensis Orphanotrophei, 1720), p.)(2r; Stahl, *Negotiosum Otiosum* in *Oeuvres, 6*, p. 3. On this debate see also L. J. Rather and J. B. Frerichs, "The Leibniz-Stahl Controversy–I, Leibniz' Opening Objections to the *Theoria medica*," *Clio medica, 3* (1968), 21–40; L. J. Rather and J. B. Frerichs, "The Leibniz-Stahl Controversy–II, Stahl's Survey of the Principal Points of Debate," *Clio medica, 5* (1970), 53–67; Paul Hoffmann, "La controverse entre Leibniz et Stahl sur la nature de l'âme, *Studies on Voltaire and the Eighteenth Century, 199* (1981), 237–241; François Duchesneau, "Leibniz et Stahl: Dívergences sur le concept d'organisme," *Stud. Leibnitiana, 27* (1995), 185–212. It should be noted that the main interests of Rather, Frerichs, and Hoffmann are on Stahl's views on the role of the *anima* in life.

[133]Stahl, *Negotiosum Otiosum* (1720), p.)(4r; *Oeuvres*, 6, p. 9.

[134]Rather and Frerichs, *op. cit.* (Part I), p. 31.

[135]*Ibid.*

[136]Stahl, *Negotiosum Otiosum* (1720), p. 13; *Oeuvres, 6*, p. 29.

[137]Stahl, *Negotiosum Otiosum* (1720), p. 52; *Oeuvres, 6*, p. 91.

medical art."[138] Further, "I assert that neither the physicians, nor the chemists, in spite of all their pretensions, have discovered anything positive touching the changes which operate so easily and diversely in the living body of man."[139]

Leibniz disagreed and stated that chemistry had indeed made progress. He asked Stahl to concede that chemistry had a great utility in the explanation of insensible phenomena that take place in living bodies. "One hopes that the discoveries in this science will become more and more frequent. Indeed, we see in animals processes that seem like chemical effervescences."[140] Leibniz continued, stating that chemistry teaches on the one hand what is common to the three kingdoms, and on the other, that which is proper to each of them in particular. For example, combustible substances such as mineral sulphur, vegetable oil, and animal fat can be equally observed in the three kingdoms, just as one can also verify the presence of volatile salt, which occurs in the animal kingdom, which one also encounters in the vegetable kingdom, and which one can even find traces of in the mineral kingdom. On the other hand, each kingdom possesses its own virtues that are particular to it. Only from the mineral kingdom can one extract the strong waters [mineral acids] and it is only from the vegetable kingdom that one obtains ardent spirits or rectified alcohol. And last, it is only from the animal kingdom that one obtains phosphorus.[141]

For Stahl this was unimpressive. He felt that Leibniz had only proved that he had some common knowledge of general chemistry.[142] But Leibniz in reply maintained "that the animal body is a veritable hydraulic-pneumatic machine and that its vehement movements have for their cause explosions like those of steam engines." Today, he added, no one doubts this except those who follow chimeric principles.[143] In short, Leibniz accepted a mechanic-hydraulic understanding of the body while approving the study of chemistry as an important aid for the physician.

---

[138]Stahl, *Negotiosum Otiosum* (1720), p. 53; *Oeuvres, 6*, p. 92.

[139]Stahl, *Negotiosum Otiosum* (1720), p. 55; *Oeuvres, 6*, p. 95.

[140]Stahl, *Negotiosum Otiosum* (1720), p. 161; *Oeuvres, 6*, p. 263.

[141]Stahl, *Negotiosum Otiosum* (1720), p. 162; *Oeuvres, 6*, p. 264.

[142]Stahl, *Negotiosum Otiosum* (1720), p. 162; *Oeuvres, 6*, p. 265.

[143]*Ibid.*

STAHL'S CHEMISTRY, ALCHEMY, AND A NOTE ON PHLOGISTON

Regardless of their rejection of the role of chemistry in medicine, it should be kept in mind that Boerhaave, Hoffmann, and Stahl were chemists as well as physicians. Leibniz also maintained a lifelong interest in chemistry.[144] Boerhaave's text became standard for mid-eighteenth-century students. Hoffmann's *Chymia rationalis et experimentalis* (1748) went through three editions, and Stahl's theory of phlogiston became a significant unifying system for pre-Lavoisier chemists.

For many chemists, a knowledge of Stahl's views came from lectures that went back to his time at Jena (1684). His *Fundamenta chymiae dogmaticae et experimentalis* (1723) derived from those lectures and appeared in almost complete form in English translation by Peter Shaw as the *Philosophical Principles of Universal Chemistry* (1730). The *Fundamenta chymiae* was to appear in other Latin editions through 1746–1747.

From the start, Stahl took up the relation of chemistry to medicine: "... *universal Chemistry* is commodiously resolved into several *Parts* or *Branches*, under which it must be distinctly treated to give a just notion of its due extent and usefulness. For tho' in common acceptation of the word, *Chemistry* is supposed to relate chiefly to the *Art of Medicine* as it supplies that Art with Remedies, this is in reality but a small part of its use, compared with the rest; numerous other Arts, Trades, and mechanical Employments, Merchandize itself, and all natural Philosophy ..."[145]

Although he did not regularly identify his sources, it is evident that Stahl was well aware of earlier writers as well as his contemporaries. Paracelsus, van Helmont, and Johann Kunckel had been read with care, and J. J. Becher had influenced him greatly. His definition of chemistry certainly reflected tradition: "Universal Chemistry is the Art of resolving *mixt*, *compound*, or *aggregate* Bodies

---

[144]For a summary of Leibniz' interest in chemistry–and especially alchemy–see Partington, *op. cit.*, *2*, p. 485.

[145]Dr. George Ernest Stahl, *Philosophical Principles of Universal Chemistry: or The Foundation of a scientifical Manner of Inquiring into and Preparing The Natural and Artificial Bodies for the Uses of Life: Both in the smaller Way of Experiment, and the larger Way of Business. Design'd as a General Introduction To the Knowledge and Practice of Artificial Philosophy: Or Genuine Chemistry in all its Branches. Drawn from the Collegium Jenense of Dr. George Ernest Stahl. By Peter Shaw M.D.* (London: John Osborn and Thomas Longman, 1730), pp. 1–2.

into their *Principles*; and of composing *such Bodies* from their *Principles*."[146] But what were these principles? The chemists spoke of salt, sulphur, and mercury or salt, oil, and spirit. Stahl believed that Thomas Willis had added phlegm and earth incorrectly.[147] But neither the Peripatetic elements nor the chemical principles could be accepted "for no one has hitherto pretended to shew that these *Principles* are specifically the same in all Bodies." Only if they are thought of as "general qualities" may they "be allow'd in *Compounds*."[148]

In the search for the first principles of matter, the work of van Helmont and Becher was considered far better. The former had taken water for the first and only material body of things, while Becher took water and earth further, distinguishing earth into three kinds, "(1) *vitrifiable* or *fusible* [terra vitrescible], (2) *inflammable* [terra pinguis] in composition, and (3) *liquifiable* or *specifically mercurial* [terra fluida] in *Metals*."[149] For Stahl the choice was clear, "we need not go so far back as to *Water* for the *Principles of Mixts*; but should rather prefer the Opinion of *Becher*, a man who seems to have been design'd for the real Improvement of *nature Philosophy*."[150]

Stahl was well aware of the long history of chemistry.[151] He offered the reader a short history of the subject going back to the Egyptians, whose wealth could best be explained by their ability to change base metals to gold and silver. So dangerous was this to the stability of the Roman Empire that Diocletian had ordered all alchemical manuscripts to be burned along with those who were skilled in the art. The Arabs, nevertheless, continued this study and Paracelsus was famous for his acceptance of transmutation, even though earlier authors such as Isaac Hollandus and Basil Valentine wrote of the art with greater depth.

This history draws attention to Stahl's persistent interest in transmutational alchemy. In this regard he raised the question of the growth of metals in the earth, a topic he had discussed also in his *Specimen Becherianum* (1703). In the *Philosophical Principles*, Stahl argued for the formation of metals in the earth from subterranean vapors. Traditionally it had been argued that such a growth process was slow. But Stahl believed that "it may be more easily conceived, that Metals should be . . . instantaneously generated, than that the

[146] *Ibid.*, p. 1.

[147] *Ibid.*, p. 4.

[148] *Ibid.*

[149] *Ibid.*, p. 8.

[150] *Ibid.*, p. 5.

[151] *Ibid.*, pp. 394–395.

imperfect Metals should . . . be converted into the perfect . . . by long continued Concoction in the Earth, or by lying therein for some hundreds of years, without the addition of any new matter, or any diminution of the old. . . . [And] if this instantaneous Generation of Metals cou'd once be well establish'd, it would give great countenance to the action of the philosophical *Tincture* or *Substance* in the business of Transmutation."[152]

Like Boerhaave, Stahl accepted the theoretical possibility of transmutation. Discussing philosophical gold, he wrote that the adepts require this "for the fermentation of their Mercury. And, indeed, they mean no more by it than gold most highly subtilized, and brought to a degree of fermentative mobility; so that being mix'd with pure running Mercury, it may by degrees assimilate the particles thereof to itself, and at length reduce the whole mass to a due degree of spissitude; whence the Mercury also may, in time, become the true *Gold*; which, tho' softer than the common, is yet like it, in sustaining all the proofs whereby the constancy of that is examin'd."[153] For those interested in investigating the matter further, he recommended the *Introitus apertus* of Eiraeneus Philalethes.[154]

Stahl also looked on the alchemist, Isaac Hollandus, as an author of primary importance in any study of transmutation. He reprinted Hollandus's treatise on the salts and oils of metals as an appendix to his own views on the subject.[155] This was included in the Latin versions of the *Fundamenta chymiae*, but not in the English translation by Shaw. Stahl also had a special interest in the mercuries of metals, which he said were used primarily for their "advancement into Gold, by a moderate digestion with highly subtilized or philosophical Gold, whence they are said to be coagulated with it into a fix'd Precipitate; which if thrown into Silver or Gold in fusion, there proves and remains good and fix'd Gold."[156]

Stahl stated that the alchemical adepts had described three basic processes leading to the preparation of the philosophers' stone by way of

---

[152]*Ibid.*, pp. 249–251. On Stahl's alchemy, see also the recent article by Johanna Geyer-Kordesch, "Chemie und Alchemie: J. J. Becher, G. E. Stahl und J. C. Carl und J. C. Dippel, *Johann Becher (1635–1682)*, ed. Gotthardt Fruhsorge und Gerhard F. Strasser (Wiesbaden: Otto Harrassewitz, 1993), pp. 127–142.

[153]Stahl, *Philosophical Principles*, p. 319.

[154]*Ibid.*, p. 321.

[155]The Hollandus treatise does not appear in the Shaw translation, but will be found in the Latin editions such as Georg Ernst Stahl, *Fundamenta Chymiae Dogmaticae et Experimentalis* . . . (Nuremberg: Imbensis B. Guolfg. Maur. Endteri, 1746), part I, pp. 237–255.

[156]Stahl, *Philosophical Principles*, p. 391.

vitriol, niter, or mercury. But he added that no matter how diligent the operator, he must realize that the ultimate result would be determined by the Divine Will and by the morality of the person: "let everyone examine himself by the Rule and accordingly expect success or failure, in his attempts."[157]

Gough has argued convincingly that Stahl and his followers made composition the basic goal of chemistry,[158] but traditionally Stahl's influence in chemistry has been associated with the phlogiston theory. Here the early lectures from Jena help us little, for it was not until Stahl had studied the work of J. J. Becher (1635–1682) that he introduced phlogiston, a word used by Aristotle in the *Meteorologica* to mean inflammable (Φλογιστοσ) and by a series of seventeenth-century authors [Hamerius Poppius (1618), van Helmont, Becher, Hapelius (1606), Sennert (1619), and Boyle (1680)].[159] Becher's second earth was also called inflammable earth or sulphur (Φλογιστον).[160] Nevertheless, these references are not numerous and we do not find any effort to establish the word or concept as the basis of a chemical system. Stahl referred to phlogiston in his treatise on fermentation, the *Zymotechnia fundamentalis* (1697), but his connection with the work of Becher is most evident in the *Specimen Becherianum* (1703). Published with Becher's *Physica subterranea* (first edition, 1669), Stahl emphasized a chemical system that remained basically inorganic, separated from life processes. A rare reference to phlogiston in regard to respiration occurs in Stahl's discussion of physiology.[161] Here he describes the process as fundamentally mechanical. Air adds little to respiration with the possible exception of phlogiston, but its quantity in the atmosphere is so small that it is unlikely that it would add anything of significance to the blood. In any case, such considerations have little value for medical practice and can be ignored.

However, it is not our purpose to discuss the phlogiston theory as a whole because Stahl employed it largely for inorganic explanations centered primarily on combustion. In this regard Stahl described many observations that today would be termed oxidation or reduction processes. Here and in his medical views, he presented systems that were in large part divorced from one another.

---

[157] *Ibid.*, p. 424; *Fundamenta Chymiae* (1746), part I, p. 236.

[158] J. B. Gough, "Lavoisier and the Fulfillment of the Stahlian Revolution," *Osiris*, *4* (1988), pp. 15–33.

[159] On pre-Stahlian uses of the word phlogiston, see Partington, *op. cit.*, *2*, pp. 667–668.

[160] *Ibid.*, *2*, pp. 645–647.

[161] *Ibid.*, *2*, p. 658; Stahl, *Oeuvres*, *3*, pp. 252–253.

CONCLUSION

Although we are accustomed to look back from the work of Lavoisier, there is much to be learned from a study of the move forward chronologically toward the late eighteenth century. As seen in this chapter, the debate over the use of chemistry continued well into the new century. In England a debate between apothecaries and physicians involved the right to prescribe and minister to patients. Here Robert Pitt attacked both the chemists and the mechanists who sought to control medicine (1704). Similarly Thomas Morgan rejected the followers of Paracelsus and van Helmont no less than Bellini and the mechanists (1735), preferring instead the work of Thomas Willis, who reflected both the recent chemistry and atomistic explanations.

The debate is also reflected in the increasing number of medical histories in this period. In the late seventeenth century Hermann Conringius and Olaus Borrichius had argued over the truths of the ancient physicians versus the Paracelsian chemists, but in the new century Daniel LeClerc and John Freind reflected more recent thought. In the second edition of his *Histoire de la médecine* (1723), LeClerc extended his discussion from the time of Galen to the middle of the seventeenth century. Here he placed emphasis on the rise of chemistry in relation to medicine and indicated the considerable influence of the work of Paracelsus. In England John Freind looked on this work with alarm. Both a physician and a chemist himself, Freind was a Newtonian who attempted to understand chemical action in terms of atoms and forces–a far cry from the views of the iatrochemists. He preferred to avoid Paracelsus entirely, arguing that the invention of any useful chemical remedies belonged to the Arabs.

In the work of the medical historians, we may see a continued awareness of debates between chemists and mechanists, but perhaps the most important early eighteenth-century authors who influenced medicine and chemistry were the contemporaries Hermann Boerhaave and Georg Ernst Stahl. Both were physicians and both became influential chemists as well. Boerhaave taught at Leiden until his death in 1738. It is evident from his earliest academic orations that he viewed the body in terms of mechanics. With the exception of Hippocrates, he attacked the ancients and the Islamic authors. The Cartesians and the chemists had added new thought, but for Boerhaave this was far less important than the work of Hippocrates. Highly critical of the chemical physicians, Boerhaave became Professor of Chemistry in 1718. It was a subject he loved. Indeed, his chemical textbook (1732) became the most popular in this field for decades. However, his presentation of chemistry was largely centered on what we would call inorganic chemistry. It was to have little effect on medicine or physiology other than in the limited preparation of pharmaceutical products and in chemical analyses. Oddly enough, he referred to van Helmont as a model for the physician and he devoted much of his time to experiments related to alchemy.

Georg Ernst Stahl, a contemporary of Boerhaave, also influenced the chemical and medical thought of his century. His development of the Phlogiston Theory was to become the predominant chemical theory prior to the work of Lavoisier and his colleagues. Like Boerhaave, he rejected chemistry as a basis for the understanding of physiology and life processes. A deeply religious man, he sought life in a controlling *anima* that directed life processes. Besides chemistry, Stahl also rejected the minutiae of the anatomists and the mechanists who sought to explain man as an automaton. Although he did not refer to Boerhaave–and Boerhaave did not refer to him–Stahl disagreed with his colleague, Friedrich Hoffmann, and with Gottfried Leibniz, regarding his vitalistic views. Both Hoffmann and Leibniz were basically mechanists and once again, both were interested in chemistry. Like Stahl, Hoffmann sought to separate chemistry and medicine, but he found considerable value in chemically prepared pharmaceuticals. Leibniz disagreed with Stahl regarding the value of anatomy and the use of chemical explanations in life processes. Both Leibniz and Stahl had an avid interest in transmutational alchemy and the latter included a work by Isaac Hollandus in his chemical text, the *Fundamenta chymiae.*

Boerhaave as a mechanist and Stahl as an animist had presented two basically different approaches to life and medical physiology. However, they both agreed that although chemistry might have some value as a servant to medicine, the two fields were quite different. This was a fundamental break with the Paracelsian-Helmontian tradition.

CHAPTER 7

Final Thoughts

[Chemistry] is so intimately connected with medicine, that to improve in the former is . . . to proceed in the latter.[1]

David Uwins, *Modern Medicine* (1808)

As far as chemistry has mingled in discussions of . . . [physiology, pathology, or therapeutics] . . . it has not only darkened them, but filled them with error.[2]

Charles Caldwell, Lecture at the Medical School in Philadelphia (1818)

In the early decades of the eighteenth century there was no real consensus among physicians regarding the correct approach to medicine. The mechanists may well have been in the ascendant, but they continued to rail at the chemists and at the same time there were those who continued to recommend Hippocrates as their primary authority.

[1]David Uwins, *Modern Medicine, Containing a Brief Exposition of the Principal Discoveries and Doctrines that Have Occasioned the Recent Advancement of Medical Philosophy, with Strictures on the Present State of Medical Practice* . . . (London: Samuel Tipper, 1808), p. 101.

[2]Charles Caldwell as cited by Herbert S. Klickstein in "Charles Caldwell and the Controversy in America Over Liebig's *Animal Chemistry*," *Chymia*, *4* (1953), pp. 129–57 (133) from T. Cooper, "Discourse on the Connection between Chemistry and Medicine, Delivered in the University of Pennsylvania, Nov. 5, 1818" (Philadelphia, 1818), p. xii.

A SAMPLER OF EARLY EIGHTEENTH-CENTURY MEDICAL HISTORY

In his *State of Physick, Ancient and Modern* (1732), Francis Clifton (d. 1734) argued that the emphasis on mathematics and the mechanical arts, beginning in the sixteenth century, had transformed the sciences as a whole, but this new emphasis had also been the reason why medicine "has been but little improv'd in comparison."[3] A basis for the needed improvement might be found in publishing accurate editions of ancient texts and by continuing anatomical research. However, Clifton warned against the excesses of the chemists, "Paracelsus at the head of 'em."[4]

Clifton's emphasis on Hippocrates was echoed by John Barker (1708–1748), an Oxford M.D. and Fellow of the Royal College of Physicians (1746). In his *Essay On the Agreement betwixt Ancient and Modern Physicians* (1746), he emphasized both the work of the ancients, Hippocrates and Galen, and the moderns, Sydenham and Boerhaave. He believed that "the best practical Writers amongst the Moderns" have embraced the Hippocratic doctrine and they "have thought it impossible to lay down a better Plan, or to establish the Practice of Physick upon a more solid, and rational Foundation."[5]

Following an historical approach, Barker insisted that Galen was a true student of nature and of Hippocrates and that while the Galenic texts had been corrected in the Renaissance, there had not been an immediate attempt to discard them. This had changed with the work of Paracelsus, who had questioned all things.[6] He, in turn, had been followed by van Helmont and his many followers. Their attempted medical revolution had far-reaching effects in spite of the absurdity of their views. And yet, Barker asserted that on many points van Helmont had really followed Hippocrates and was not as revolutionary as he seemed. Similarly, although the mechanists may have

---

[3]Francis Clifton, *The State of Physick, Ancient and Modern, Briefly Consider'd: With A Plan for the Improvement of it* (London: W. Bowyer for John Nourse, 1732), p. 117. Clifton received his M.D. at Leiden in 1724 and was elected a Fellow of the Royal College of Physicians. He was physician to the Prince of Wales and wrote a number of medical tracts before moving to Jamaica in 1734. His *State of Physick* was translated into French in 1742.

[4]*Ibid.*

[5]John Barker, M.D., *An Essay On the Agreement betwixt Ancient and Modern Physicians: Or a Comparison Between the Practice of Hippocrates, Galen; Sydenham and Boerhaave in Acute Diseases. Intended to shew, What the Practice of Physick, in such Distempers, ought to be* (London: G. Hawkins, 1747), p. 162.

[6]*Ibid.*, p. 164.

improved on some aspects of the Hippocratic tradition, both Boerhaave and Hoffmann had been followers of Hippocrates.

A few years earlier Ashley Cowper (d. 1788), writing as Timothy Scribble, wrote a poetical *Progress of Physick* (1743) in which he pointed to the revival of medicine by the Arabs, whose labors were vitiated by their acceptance of Galen's errors.

> Next *Chymistry*, which long in *Embrio* lay,
> Started new Lights . . . and smooth'd the thorny Way;
> Nor knew t'*enrich*, alone, but *bless* Mankind.
> Long had she learn't t'extract the shining, *Ore*,
> But Med'cine now confess'd her *healing Pow'r*.
> A new *Creation* open'd to the View,
> And old Discov'ries were confirm'd by new.[7]

Cowper noted that Paracelsus and van Helmont carried their work in chemistry to such a length "as to render Medicine almost wholly Chymical," but understood that Robert Boyle had "rescued *Chemistry* from the censure it had long lain under from the *Enthusiasm* of *Helmont* and *Paracelsus*."[8] Indeed, medicine had of late "become all *Mechanical*, and *Corpuscular*."[9]

CHEMICAL MEDICINE FROM PARACELSUS TO BOERHAAVE

The works of Clifton, Barker, and Cowper were not meant to be major histories and it would be wrong to compare them with the efforts of Daniel LeClerc or John Freind. Nevertheless, they do indicate that chemistry remained a subject of contention in medical circles in the first half of the eighteenth century.

As we have seen, the literary Renaissance of the fifteenth century was reflected in medicine in the sixteenth century in new translations of ancient authorities that led to a dominance of Galen in the medical pantheon. Chem-

[7]Ashley Cowper, *The Progress of Physick, by Timothy Scribble, Esq.* (London: J. Stagg for J. Roberts, 1743), p. 18.

[8]*Ibid.*, p. 20.

[9]*Ibid.*, p. 19.

istry played little part in this movement, but a real change came with the Swiss-German physician, Paracelsus. A rebel who opposed establishment medicine, he shocked the doctors at Basel by throwing the *Canon* of Avicenna on the St. John's Day bonfire in 1527. His contempt of the universities was to become a hallmark of his sixteenth-century followers. Indeed, Paracelsus had sought a new philosophy of nature based primarily on chemistry and chemical analogies. His was a universe of macrocosm and microcosm bound together in a single whole. He rejected the humors of the Galenists and sought to explain the body in terms of chemistry or chemical analogies.

All of this was anathema to the medical establishment. But there was more. The Paracelsians insisted that like cures like, that a poison in the body would be cured by a similar poison. This was in direct contrast to the Galenic belief that contraries cure. They accused these chemical physicians of murdering their patients. For many this charge seemed difficult to dispute, since a major innovation of the chemists had been to introduce remedies prepared from metals, far stronger in their purgative and vomitive effects than the traditional Galenicals.

The debate between the chemical physicians and the Galenists was both inevitable and bitter. One point of contention was the use of the new chemical remedies, which may best be seen in the so-called "antimony war" that affected all European medical circles for a century. And yet, by the third quarter of the seventeenth century many of these chemically prepared drugs were accepted by physicians and used along with the traditional materia medica. Similarly, the medical faculties of European universities began to establish chairs in chemistry.

As the chemically prepared remedies were gradually being accepted by the medical community, there was a new interest in understanding the living body in chemical terms. Here van Helmont's work was of special influence. Interested in a new chemical philosophy of nature, he gave new life, and to a degree, a new direction to Paracelsism. His work on digestion, fermentation, and acid-alkali reactions spawned a new generation of chemical physicians while sharpening the attacks of their opponents.

Prominent among mid-seventeenth-century iatrochemists were Thomas Willis and Franciscus de la Boë Sylvius, both famous anatomists who sought to understand disease and the function of the organs of the body through chemistry. In England there is no doubt that Robert Boyle was strongly influenced as a young scholar by the work of van Helmont, while Thomas Sherley tried to show the similarities between traditional and chemical medicine. But confrontation was more common, as we have seen in the failed attempt to establish a chemical society of physicians in London in 1665.

Also influenced by the Helmontian corpus was Otto Tachenius, who sought to ascribe natural philosophy–which to him could be equated with chemistry–to Hippocrates. The works of Tachenius, short though they were, were frequently reprinted and became the basis of debate well into the next century.

The seventeenth century saw a gradual decline in confidence in the ancient authors other than Hippocrates. In part this may have been due to the attacks of the chemists, but there is little doubt that the development of the mechanical philosophy, with its emphasis on mathematics, was fully as important. The work of Copernicus, Tycho, Kepler, Galileo, and Newton established the heliocentric system as well as classical mechanics, and along with Descartes emphasized the power of mathematical proof in the physical sciences.

Hoping to emulate the works of the physicists and astronomers, some physicians sought to make medicine a mathematical and mechanical science. Borelli, Bellini, and Baglivi turned from the chemical explanations of the Helmontians to describe the body in terms of simple machines and to mathematicize the flow of liquids throughout the body. Some followed Descartes in comparing man to an automaton. Again digestion was a subject of debate. Was it due to the action of acids or perhaps fermentation as the chemists held, or was it due to the grinding action of the muscles of the stomach lining as the mechanists insisted?

By the early decades of the eighteenth century, the chemical tradition in medicine was dealt blows from two leading physicians in Europe, Hermann Boerhaave and Georg Ernst Stahl. Both were chemists: Boerhaave was famous for his chemical textbook and Stahl was responsible for his development of the phlogiston theory. However, both were also known for their medical work. Boerhaave rejected the far-reaching claims of the iatrochemists and sought to divorce chemistry from medicine, other than as an aid to the study of fluids and other analytical purposes. Although opposed to much of the work of Descartes, he did view man in mechanical terms.

In contrast to Boerhaave, Stahl sought to explain life in terms of an *anima*, a life force. Stahl also argued that chemistry was not the answer to physiology. His vitalism stood in stark contrast to the mechanism of Boerhaave. And yet, both were agreed that the chemical tradition, whether viewed in terms of the Paracelsians, the Helmontians, or the more recent followers of Willis and Sylvius, could not be accepted.

CHEMISTRY AND MEDICINE AT MID-CENTURY: THE STAHLIANS

There were some who thought that the work and the methods of both the mechanists and the chemists could be used together. Jean Baptiste Senac (1693–1770), physician to Louis XV, wrote a *Nouveau Cours de Chymie Suivant les Principes de Newton & de Sthall* (1723, reprinted 1737). Here he discussed the introduction of chemistry to medicine by Basil Valentine, Paracelsus, Boyle, Tachenius, and Glauber[10] and then went on to charge Willis with making an alembic of the human body.[11] In contrast, Pitcairne, Baglivi, and Boerhaave had brought mechanism into medicine,[12] but Senac felt that both the chemists and the mechanists had gone too far. True, he wrote, there is a value in the chemical preparation of some medicines, but the physiological processes of the human body are not to be explained by chemistry.[13] Like Newton, Senac sought to discover general laws and a force that holds bodies together. However, he felt that although many sought to apply Newtonian principles to chemistry, their results to date had been disappointing.[14]

Johann Juncker (1679–1759) praised the work of Boerhaave, but he was surely a follower of Stahl. Juncker taught at Halle in 1701–1702, took his M.D. at Erfurt in 1717, and became professor of medicine at Halle in later life. In his *Conspectus Chemiae Theoretico-Practicae . . . e Dogmatibus Becheri et Stahlii* (1730), he discussed the relation of chemistry to medicine. He was certainly more interested in the chemical understanding of the body than Stahl was, for he complained of those physiologists who tried to explain phenomena without a knowledge of chemistry. "How can one explain the process of digestion which operates in the stomach through fermentation, or the separation of the chyle," or other physiological phenomena without a knowledge of chemistry? Indeed, foods change their nature by a chemical operation, and yet, those who depend solely on an acid-alkali dichotomy are

---

[10]Jean Baptiste Senac, *Nouveau Cours de Chymie Suivant les Principes de Newton & de Sthall* (Paris: Jacques Vincent, 1737), p. lviii.

[11]*Ibid.*, p. lvii.

[12]*Ibid.*

[13]*Ibid.*, p. lxviii.

[14]See the review of Senac's book in *Le Journal de Scavans* (1724), pp. 30–34 (32). On the Newtonian impact on chemistry, see Arnold Thackray, *Atoms and Powers: An Essay on Newtonian Matter-Theory and the Development of Chemistry* (Cambridge: Harvard University Press, 1970). Senac's authorship of the *Nouveau Cours* has been questioned. Thackray discusses this literature in *Atoms and Powers*, p. 94n.

wrong, because this system collapsed "when chemists have shown that it is impossible to find in living bodies any acid, alkali, or neutral salt capable of producing the effects attributed to them."[15]

Juncker questioned the value of mathematics for a chemist. He felt that it was impossible to determine the geometric shapes of the particles of bodies and although mathematicians insist on an infinite division of matter, the chemists could not agree. Nevertheless, mathematics can be a great help in explaining the relationship of principles in regard to their qualities and weights. "The excellent work of chemists following the principles of Newton and Stahl is a model of perfection of this sort and one can see that in this respect that mathematics can shed light on chemistry."[16]

Juncker accepted the three Paracelsian principles, Salt, Sulphur, and Mercury, plus the passive principles of phlegm and *terra morte.* These five were part of the seventeenth-century chemical tradition and it is not unexpected to find Juncker also discussing alchemical transmutation at length. Even Stahl had defended alchemy and it seemed necessary to Juncker to carefully separate the true from the false in the alchemical corpus.[17]

He took aim at van Helmont and Willis, who imagined that they felt heat and saw effervescence when it did not exist. These two had placed particular ferments in different places in the human body and thought that all our humors are agitated by reason of the leaven. "Let us be assured that this quantity of different ferments is the fruit of an imagination furiously fermenting."[18]

A connection between Stahlian vitalism and chemical tradition was seen by Sieur Raguet (1758). He believed that the Spirit of Life is hidden in different individuals in nature, but it is the same everywhere and only requires its apparent difference through the subject it inhabits. "This spirit is called phlogiston in chemistry, the universal spirit which plays such a great role in nature."[19] But one could go back farther. This is the same Spirit of Life "that van Helmont has vaunted, which he presented to us under the name of Archeus,

[15]I have used the French edition of Juncker's *Conspectus Chemiae: Élémens de Chymie, suivant les Principes de Becker & de Stahl*, trans. from the Latin of the second edition of M. Juncker with notes by M. De Machy, Apothecaire Gagnant-Maîtrise de l'Hotel-Dieu de Paris (6 vols., Paris: Siméon-Prosper Hardy, 1757), *1*, pp. 15, 17.

[16]*Ibid.*

[17]*Ibid.*, *2*, chapters 18–19. On transmutation, see pp. 543, 583–589.

[18]*Ibid.*, *6*, p. 20.

[19]Sieur Raguet, *Dissertation sur Diverses Maladies et sur l'Usage de l'Esprit Philosophique; Avec un Recueil des principales guérisons opérées par ce Reméde* (Paris: Aux dépens du Sieur Raguet, Auteur de l'Esprit Philosophique, 1758), p. xi^{r}.

and which he regarded... as the origin of maladies and of health.... This great man recognized that this spirit sends other spirits emanating from it throughout our viscera."[20]

Nevertheless, Raguet was an exception to the rule in his adherence to chemistry and life. The most enduring medical influence of Stahl was to be found at the medical school of Montpellier. There François Bossier de Lacroix Sauvages (1706–1767), a follower of Baglivi, began teaching a mechanistic medicine, but soon found this unsatisfactory for understanding the processes of living matter. Turning to the work of Stahl, he began a series of lectures that introduced the *anima* to his students in 1737.[21]

Sauvages was followed by Théophile de Bordeu (1722–1776) and Paul-Joseph Barthez (1734–1806). With these figures, we go well into the second half of the century, but their connection with Stahl is clear. Bordeu was convinced that Stahl's views on physiology were based on those of van Helmont and he argued in favor of vital forces in each organ.[22] Here there is a clear similarity with the Helmontian archei. Barthez, on the other hand, preferred a single vital spirit. The many life forces in individual organs described by Bordeu seemed quite unnecessary for him.[23]

If Bordeu and Barthez disagreed somewhat regarding the *anima*, they were in complete agreement with Stahl in their belief that chemistry had little place in medical theory. To be sure, the iatrophysicists were wrong in their emphasis on mechanistic explanations, but the chemists were no better. Bordeu wrote that "The anatomists have dissected the body down to its infinitely small fibrils and the physicians have transformed man into a machine of levers, pumps, springs, pipes and presses. But the school of Paracelsus would make the body a composite of alembics, ferments, salts, effervescences, distillatory vessels and centers of explosions."[24] Barthez went on to insist, along

---

[20]*Ibid.*, 19.

[21]Elizabeth Haigh discusses the vitalistic origins of the medical school at Montpellier in her *Xavier Bichat and the Medical Theory of the Eighteenth Century, Medical History*, Supplement No. 4 (London: Wellcome Institute for the History of Medicine, 1984). See her discussion of Sauvages, pp. 28–31. See also Lester S. King, "Boissier de Sauvages and 18th Century Nosology," *Bulletin of the History of Medicine*, *40* (1966), 43–51.

[22]*Ibid.*, p. 36. See also Théophile de Bordeu, "Recherches sur l'Histoire de la Médecine" in *Oeuvres complètes de Bordeu, Précédées d'une Notice sur la vie et sur ses Ouvrages par le Chevalier Richerand* (2 vols., Paris: Caille et Ravier Libraires, 1818), *2*, p. 671.

[23]P. J. Barthez, *Nouveau Éléments de la Science de l'Homme*, Third edition (2 vols., Paris: Germer Baillière, 1858), *1*, pp. 24, 93–94.

[24]de Bordeu, "Analyse Médicinale du Sang" in *Oeuvres* (1818), *2*, p. 930.

with Stahl, that an inseparable gulf separates living and dead matter. He wrote that the vital principle is necessary for the functions of life, which are totally different from the phenomena of "dead" nature which are determined by the operations of chemistry.[25]

At Montpellier the division between chemistry and medicine was established by Sauvages, Bordeu, and Barthez. They were joined by Xavier Bichat (1771–1802), who also insisted that physics and chemistry were distinct from physiology.[26] And his authority assured that Montpellier maintained a nonchemical and vitalistic medicine throughout the nineteenth century.

A NEW DIRECTION

Surely not everyone was satisfied with the separation of chemistry and medicine seen in the works of Boerhaave and Stahl. And there were those who sought a new beginning for chemistry. The article on "chymie" in the highly influential *Encyclopédie* (1753) was written by Gabriel François Venel (1723–1759), who was later (1759) to be given the chair of chemistry at Montpellier.[27] In his article he complained that "chemistry is little cultivated among us . . . The chemists form a small number, a distinct group having their own language, their laws and mysteries, and living nearly isolated in the midst of a large population little interested in its commerce and industry."[28]

Dissatisfied with much of the literature, Venel called for a new Paracelsus to remedy the situation.[29] For him "Paracelsus was one of the most sin-

---

[25]Barthez (1858), *1*, p. 37. "Les affections du Principe Vital qui produisent et renouvellent, dans un ordre constant, les fonctions necessaires à la vie, sont absolument différentes des causes productives des mouvements qui ont lieu dans la Nature morte, comme sont ceux qui determinent les opérations de la Chimie."

[26]Haigh (1984), p. 15; M. F. Bérard, *Doctrine Médicale de l'École de Montpellier, et Comparison de ses Principes avec ceux des autres Écoles d'Europe* (Montpellier: Librairie au Rabais, 1819), pp. 311–312.

[27]G. F. Venel, "Chymie" in *Encyclopédie, ou Dictionnaire Raissonné des Sciences, des Arts et des Métiers,* edited by M. Diderot and M. D'Alembert, vol. 3 (Paris: Briasson, David, Le Breton and Durand, 1753), p. 408. Diderot's antimechanistic views have been noted by Charles Coulston Gillispie in his "The *Encyclopédie* and the Jacobin Philosophy of Science" in *Critical Problems in the History of Science,* edited by Marshall Clagett (Madison: University of Wisconsin Press, 1962), pp. 255–289. See also Gillispie, *The Edge of Objectivity* (Princeton: Princeton University Press, 1960), pp. 184–187 and Partington, *op. cit.,* vol. 3, pp. 78–79.

[28]Venel, *Encyclopédie, 3,* p. 408.

[29]*Ibid.,* p. 410.

gular personnages in literary history." On the one hand he was a visionary fascinated by astrology, the cabala, magic and all of the occult sciences, but he was bold, enthusiastic and extraordinary in all things. He was surely arrogant in giving to himself the title Prince of Medicine and Monarch of Arcana. But even if we allow for his faults, we must admit that he was the author of the greatest revolution that has ever changed the face of medicine. Although his work is "absolutely unintelligible we may credit him with the propagation and perpetuation of chemistry."[30] Here Venel referred specifically to the influence of Paracelsus on the development of chemically prepared medicines.

Venel may well have been the last major figure to turn to Paracelsus for inspiration, but his call for a new chemistry was met in the final decades of the century with the work of Antoine Lavoisier (1743–1794) and his colleagues.[31] It is not our intention to discuss here the steps leading to this development, but once again we must return to the previous century. Essential for the work of Lavoisier and his contemporaries was an understanding of the atmosphere. Van Helmont had collected various "airs" in the first half of the seventeenth century and he had been followed by John Mayow (1641–1679), who, building on a Paracelsian tradition,[32] had postulated that a "nitrous" part of the air (nitroaerial particles) was essential for combustion and respiration.

In his *Vegetable Staticks* (1727), Stephen Hales repeated an experiment by Mayow in which he placed a rat in a bell jar over water and found that a portion of the air (1/27th part) was absorbed. There was no suggestion that this portion was a distinct part of the air. In succeeding years, Joseph Black showed that "fixed air" (our carbon dioxide) combined with lime (calcium oxide) and magnesium oxide, and Joseph Priestley and Henry Cavendish conducted experiments that resulted in numerous "airs" that differed in their properties. Among others, Priestley isolated nitrogen and oxygen. However, he held to the phlogiston theory and named them phlogisticated and dephlogisticated airs. Like others before him, Priestley was interested in respiration and combustion, both of which he found to be dependent on dephlogisticated air.

---

30 *Ibid.*, pp. 431–432.

31 The literature on the Lavoisier revolution in chemistry is enormous. Of special significance for this study is Frederic Lawrence Holmes, *Lavoisier and the Chemistry of Life: An Exploration of Scientific Creativity* (Madison: The University of Wisconsin Press, 1985).

32 Allen G. Debus, "The Paracelsian Aerial Niter," *Isis*, *55* (1964), 43–61; Debus, *The Chemical Philosophy: Paracelsian Science and Medicine in the Sixteenth and Seventeenth Centuries*, 2 vols. (New York: Science History Publications, 1977), *2*, pp. 492–496.

Lavoisier followed Priestley's experiments and was soon able (1776) to show that dephlogisticated air is absorbed by metals. When this occurred, the remaining "air" was "mephitic air" (our nitrogen) that no longer supported either combustion or respiration. However, when the original vital air (dephlogisticated air) was added to the mephitic air, normal atmospheric air was restored. The end result of this and other experiments was summarized in his antiphlogistic theory,[33] which was described by Lavoisier in his *Traité Élémentaire de Chimie* (1789) and was rapidly adopted by European chemists.

The resultant understanding of respiration was accompanied by an understanding of the relation of plants to the atmosphere and the carbon cycle.[34] Presented originally in terms of Phlogiston by Jean Senebier (1742–1808), it was developed further and explained in the framework of the new French chemistry by Jan Ingenhousz (1730–1799).

And while the relationship of life and the atmosphere was a subject of investigation, there remained an interest in digestion. René Antoine Ferchault, Sieur de Reamur (1683–1757), studied the solvent power of the gastric juice of birds of prey, which turned blue paper red (1752).[35] To be sure, van Helmont had stated that this juice is acid more than a century earlier, but as we have seen, the significance of this observation had not been universally accepted. Lazzaro Spallanzani (1729–1799) reported on an analysis of gastric juice (1780) made by Scopoli, but he did not believe it was an acid and he did not think that digestion was a fermentative process.[36] In short, an understanding of the digestive process remained a subject of debate until well into the nineteenth century, when William Prout (1785–1850) proved that the acid in the gastric juice is hydrochloric acid (1824).[37] Following this, the American Army Surgeon, William Beaumont (1785–1853), published the results of his

---

[33]Conveniently summarized by J. R. Partington in his still useful *A Short History of Chemistry* (London: Macmillan, 1951), pp. 128ff. For recent interpretations, see the essays included in *Osiris*, Second Series, vol. 4, "The Chemical Revolution: Essays in Reinterpretation," edited by Arthur Donovan (1988) and Frederic Lawrence Holmes, *Eighteenth-Century Chemistry as an Investigative Enterprise* (Berkeley: University of California: Office for History of Science and Technology, 1989).

[34]In addition to Holmes, *Lavoisier and the Chemistry of Life*, see Leonard K. Nash, *Plants and the Atmosphere*, Case 5, Harvard Case Histories in Experimental Science, James Bryant Conant, General Editor (Cambridge, Mass.: Harvard University Press, 1952).

[35]Partington, *History of Chemistry*, *3*, p. 62.

[36]*Ibid.*

[37]Fielding H. Garrison, *An Introduction to the History of Medicine with Medical Chronology, Suggestions for Study and Bibliographic Data* (Fourth edition, Philadelphia and London: W. B. Saunders Co., 1961), 476 here citing Prout's paper in the *Philosophical Transactions of the Royal Society* (1824), pp. 45–49.

experiments on Alexis St. Martin (1833), who had suffered a gunshot wound to the stomach that healed with a permanent fistula and made it possible for Beaumont to observe the digestive process. He confirmed the fact that the gastric acid was hydrochloric plus another active substance that Theodor Schwann showed to be pepsin in 1835. And finally, Justus Liebig (1803–1873), who believed in a vital spirit, described his research in physiological chemistry in his *Animal Chemistry* (1842). Here he attempted to show how the materials of the body are formed chemically.

At this point we are well beyond the closing date chosen for this study. In fact, we are at the verge of the birth of modern physiological chemistry. Certainly it was becoming ever more evident that chemistry was essential for an understanding of life processes. Among major medical schools only Montpellier held to the vitalist views of its eighteenth-century physicians. And yet, even there, Isaac Gaussen spoke on his *Réflexions sur l'Application de la Chimie a la Medecine* (1798). Here he compared the low state of medical knowledge with that of the other sciences and concluded that chemistry was not used often enough or well enough in medicine.[38]

Perhaps a fitting conclusion to the impact of the new chemistry on medicine in the early years of the new century may be seen in a work by David Uwins (1780–1837), a fellow of the Royal College of Physicians of London (1807). In his *Modern Medicine: Containing a Brief Exposition of the Principal Discoveries and Doctrines that Have Occasioned the Recent Advancement of Medical Philosophy* (1808), he summarized the history of medicine. Here he pointed out that "with Paracelsus and all of his sect, all was chemistry, but a chemistry widely different, in its first principles and in its practical application, from that which in the present day bears this appellation."[39] Uwins attacked the views of van Helmont and Stahl and then went on question the mechanists. Almost in despair, he noted that "the human body has now been converted into a chemical alembic, now into a hydraulic machine, its functions have been explained at one time by numbers, at another by forms, and at another by figures . . ."[40]

---

[38]Isaac Gaussen, *Réflexions sur l'Application de la Chimie a la Médecine. Présentées à l'École de Médecine de Montpellier, le quatre Thermidor, au six de la République* (Montpellier: Jean Martel, VI (1798), pp. 6–7.

[39]Uwins (1808), p. 17. See above, note 1. On his life and career, see William Munk, *The Roll of the Royal College of Physicians of London* . . . (3 vols., 2nd edition, London: Published by the College, 1878), *3* (1801–1825), pp. 56–58. Uwins took his M.D. degree at Edinburgh in 1803 and wrote six books on medicine.

[40]*Ibid.*, p. 43.

In a chapter on chemistry, he stated bluntly that chemistry and medicine "are absolutely inseparable."[41] He turned to the eighteenth-century developments in chemistry, particularly those of Joseph Black and Antoine Lavoisier. He went on to discuss the new understanding of the atmosphere and its relation to respiration, which he felt had clearly been shown to be a chemical process.[42] Digestion should be understood primarily as a chemical process because "the assimilation of the food consists merely in a certain number of chemical decompositions which the food undergoes . . ."[43] Uwins added that all must agree that in addition to a greater understanding of physiology, chemistry offered the physician numerous valuable medicines.

A century before Uwins, many physicians believed that no one could go beyond the work of Hippocrates. But Uwins wrote that because of recent discoveries, modern medicine surpassed that of the ancients. "I believe there is no man, however sceptical, who would at this moment trust himself in the hands of a physician equal in sagacity and judgment to Hippocrates, with only Hippocrates' knowledge."[44] And much of this knowledge, he felt, was due to the advances in chemistry, because chemistry "is so intimately connected with medicine, that to improve in the former is . . . to proceed in the latter."[45]

POSTSCRIPT

The controversies since the time of Paracelsus concerning the place of chemistry in medicine certainly did not end with the mechanists and the vitalists in the eighteenth century. And although the work of Lavoisier and his colleagues led the way to a new understanding of chemical physiology while a series of pharmacist-chemists successfully sought to isolate the active ingredients of medicinal substances, there remained those who rejected any significant role of chemistry in medicine.

In 1953 Herbert S. Klickstein examined the American controversy over Justus Liebig's *Animal Chemistry* after its translation into English and publication in the United States (1842).[46] At the medical school at the University of

---

[41] *Ibid.*, p. 80.

[42] *Ibid.*, pp. 121–125.

[43] *Ibid.*, p. 128.

[44] *Ibid.*, pp. 189, 192.

[45] *Ibid.*, p. 101.

[46] Klickstein, *Chymia* (1953), 133 (see note 2 above).

Pennsylvania, the professor of chemistry Robert Hare had been forced to give up his right to examine medical students at their final examination in any field other than chemistry (1818). The chief opponent to chemistry was Charles Caldwell (d. 1853), then professor of natural history, who gave a lecture to medical students in 1818 in which he attacked the use of chemistry in medicine.

> To some of you, it may, perhaps, appear extraordinary, that I have represented chemistry, as only a collateral element of medicine. . . .
>
> Shall I be told that chemistry aids, in the explication of any of the phenomena or laws of the living body, either in a healthy or a diseased state?–that it sheds light on physiology, pathology, or therapeutics? From the most correct and satisfactory views I have been able to form on this subject, I feel myself compelled to deny the position. As far as chemistry has mingled in discussions of this nature, it has not only darkened them, but filled them with error. It has superadded corruption to what it found already sufficiently corrupt.[47]

These views could well have been written at least a century earlier. Moving to Kentucky a decade later, Caldwell continued his attack on the chemical physiologists. He rejected the chemical views of William Beaumont in his review of *Experiments and Observations on the Gastric Juice, and the Physiology of Digestion* (1834), and produced a pamphlet against Liebig's *Animal Chemistry* in 1843. Here Caldwell again condemned chemical physiology as physically groundless and morally pernicious.[48] Caldwell's opposition continued until the time of his death. In 1852 he wrote once more against Liebig in a pamphlet on vital chemistry. Here he did not "hesitate to pronounce the chemical dogmas of the celebrated Professor of Giessen, when applied to living organic matter, whether vegetable or animal, as groundless and untenable as the wildest vagaries that ever issued from the tongue or pen of that prince of boasters, fictionists and notion-mongers, the renowned AUREOLUS PHILLIPPUS THEOPHRASTUS BOMBASTUS DE HOHENHEIM PARACELSUS."[49]

Caldwell may not have been the last medical vitalist, but he surely was one of the most vocal in a period when the significance of chemistry was

---

[47] *Ibid.*, 133.

[48] *Ibid.*, 147.

[49] *Ibid.*, 156.

gaining ever more adherents. And yet, even today there are those in the medical community who deplore the required chemistry courses in their training. Only a month prior to writing these lines I spoke on the long history of debate that had resulted from the introduction of chemistry to medicine. I was not too surprised to find a senior member of the medical community rise to complain of the waste of time the courses in chemistry he had taken had been. Indeed, he implied that chemistry could well be eliminated from the medical curriculum.

Bibliography

This bibliography is divided into primary and secondary sources, with the works of individual authors arranged chronologically. Titles of books in foreign languages have been transcribed as they appear on the title pages regardless of modern conventions relating to diacritical marks. I have limited myself here to works actually cited in the text, although other works have been consulted. As a result, this bibliography should not be considered exhaustive in scope for the subject.

As always, J. R. Partington's *History of Chemistry* and Lynn Thorndike's *History of Magic and Experimental Science* have proved invaluable for identifying authors and specific works. Many of the titles referred to are very rare and I wish to acknowledge again my debt to the great collections of the Wellcome Historical Medical Library and the British Library in London, the National Library of Medicine in Bethesda–and the Duveen and Cole Collections at the University of Wisconsin in Madison, where John Neu has almost acted as a "collaborator" when I have visited there. However, all of these libraries are very strong in early medical and chemical works and in every case the members of their staffs have been most helpful.

Although not a major figure in my text, special mention should be made here of the current interest in the work of Robert Boyle. The new edition of his work appeared after I had written the section on him in Chapter 3. As a result, my references there refer to the 1772 London edition of his *Works*. However, I have noted a number of specific studies by authors who have been reassessing his body of work and these studies have been cited here.

A. PRIMARY SOURCES

Académie Royale des Sciences (Paris). *Histoire de l'Académie Royale des Sciences . . . Avec les Mémoires de Mathématique & de Physique . . . Tirées des Registres de cette Académie*, from 1699. I have used the 2nd Amsterdam edition published by Pierre Mortier and others (from 1734).

Aignan, François, Medecin du Roy, & Docteur de la Faculté de Padouë. *L'Ancienne Medecine a la Mode, ou le Sentiment Uniforme d'Hippocrate & de Galien sur les Acides & les Alkalis.* Paris: Laurent d'Houry, 1693.

Aignan, François, Medecin du Roy, & de M. le Cardinal de Furstemberg & Docteur de la Faculté de Padouë. *Le Prestre Medecin ou Discours Physique Sur l'établissement de la Medecine. Avec un Traite du Caffé & du Thé de France, selon de Systeme d'Hippocrate.* Paris: Laurent d'Houry, 1696.

André, Fran., Dr. in Physick, of the Faculty of Caën. *Chymical Disceptations: Or DISCOURSES upon ACID and ALKALI. Wherein are Examined the OBJECTIONS of Mr. Boyle against these PRINCIPLES. Together with a REPLY to a Letter of Mr. S: Doctor of Physick, & Fellow of the Colledg of ***, wherein many Errors are corrected, touching the Nature of these two SALTS, Faithfully rendred out of French into English by J. W. To which is added, by the Translator, a Discourse of Phlebotomy, shewing, the Absolute Evils, together with the Accidental Benefits thereof, in some cases.* London: Printed by Tho. Dawks and Benj. Allport, 1689.

Angeleres, Padre Fr. Buenaventura Angel, Real Filosofia, Vida de la Salvd Temporal Sabidvria Sophica. *Testamento Filomedico Arcanos Filochimicos Hipocratica, Galenica, Lilibetonica. . . .* Madrid: D. Mariana del Valle, 1692.

Astruc, Jean. *Traité de la cause de la Digestion ou l'on refute le nouveau Sistéme de la Trituration & du broïement. Et où l'on prouve que les alimens sont digerez & convertis en chile, par une veritable Fermentation.* Toulouse: Ant. Colomiez, 1714.

Astruc, Jean. *Epistolae Joan. Astruc, quibus respondetur epistolari dissertatione Thomae Boeri de concoctione.* Toulouse, 1715.

Bacon, Roger. *The Opus Majus of Roger Bacon.* Translated by Robert Belle Burke. 2 vols. New York: Russell and Russell, Inc., 1962. Reprint of 1928 edition.

Baglivi, Giorgio. *The Practice of Physick, Reduc'd to the Ancient Way of Observations Containing a Just Parallel between The Wisdom and Experience of the Ancients, and The Hypothesis's of Modern Physicians. . . .* Translated by George Sewell and J. T. Desaguliers. 2nd edition, London: Andr. Bell et al., 1704.

Barker, John. *An Essay On the Agreement betwixt Ancient and Modern Physicians: Or a Comparison Between the Practice of Hippocrates, Galen; Sydenham and Boerhaave in Acute Diseases. Intended to shew, What the Practice of Physick, in such Distempers, ought to be.* London: G. Hawkins, 1747.

Barthez, P. J. *Nouveau Élemens de la Science de l'Homme.* 3rd edition. 2 vols. Paris: Germer Baillière, 1858.

Basil Valentine. *The Last Will and Testament of Basil Valentine, Monke of the Order of St. Bennet . . . To which is added Two Treatises, the First declaring his Manual Operations. The Second shewing things Natural and Supernatural. . . .* Translated by J(ohn) W(ebster). London: S.G. and B.G. for Edward Brewster, 1671.

Bérard, M. F. *Doctrine Médicale de l'École de Montpellier, et Comparison de ses Principes avec des autres Écoles d'Europe.* Montpellier: Librairie au Rabais, 1819.

Bertrand, Mr., Docteur en Medecine Aggregé au College des Medecins, de Marseille. *Reflexions nouvelles sur l'Acide et sur l'Alcali: Où apres avoir demontré que ces deux sels ne peuvent pas être les principes des Mixtes, On fait voir le veritable usage qu'on en peut faire dans la Physique & dans la Medecine.* Lyon: Thomas Amaulry, 1683.

Boerhaave, Hermann. *A New Method of Chemistry: Including the Theory and Practice of that Art: Laid Down on Mechanical Principles, and accomodated to the Uses of Life. The Whole making a Clear and Rational System of Chemical Philosophy. To which is prefixed A Critical History of Chemistry and Chemists, From the Origin of the Art to the present Time,* trans. P. Shaw and E. Chambers. London: J. Osborn and T. Longman, 1727.

Boerhaave, Hermann. *A New Method of Chemistry: Including the History, Theory, and Practice of the Art, Translated from the Original Latin of Dr. Boerhaave's Elementa Chemiae, As Published by Himself. To which are added Notes; and an Appendix, Shewing The Necessity and Utility of Enlarging the Bounds of Chemistry. With Sculptures.* By Peter Shaw. 2nd edition. 2 vols. London: T. Longman, 1741.

Boerhaave, Hermann. "De mercurio experimenta." *Philosophical Transactions of the Royal Society, 38* (1733–1734), 145–167; *39* (1735–1736), 343–359, 368–376. The first of these appeared in English as *Some Experiments Concerning Mercury . . . Translated from the Latin, Communicated by the Author . . . to the ROYAL SOCIETY.* London: J. Roberts, 1734. A French version of the first article will be found in the *Mémoires . . . de la Academie Royale des Sciences,* 1734; Amsterdam: Pierre Mortier, 1738, 739–59.

Boerhaave, Hermann. *Academical Lectures on the Theory of Physic. Being a Genuine Translation of his Institutes and Explanatory Comment . . .* 6 vols. London: W. Innys, 1742–1746.

Boerhaave, Hermann. *An Essay on the Virtue and Efficient Cause of Magneticall Cures . . .* London: n.p., 1743.

Boerhaave, Herman. *Boerhaave's Correspondence.* Edited by G. A. Lindeboom. 3 vols. Leiden: E. J. Brill, 1962, 1964, 1979.

Boerhaave, Hermann. *Boerhaave's Orations.* Translated with introductions and notes by E. Kegal-Brinkgreve and A. M. Luyendijk-Elshout. Leiden: E. J. Brill/Leiden University Press for the Sir Thomas Browne Institute, 1983.

Boix y Moliner, Doct. D. Miguel Marcelino. *Hippocrates Defendido, de las Imposturas, y Calumnias, que algunos Medicos poco cautos le imputan: En particular en la Curacion de las enfermedades agudas: pues hasta aora todavia se ignora como las curava; con sola la Exposicion, ò comento del primer Aphorismo: Vita brevis, Ars vero longa, &c.* Madrid: Matheo Blanco, 1711.

Boix y Moliner, Miguel Marcelino. *Hipócrates Defendido.* Mexico, D.F.: Imp. Del Sagrado Corazón de Jesús, 1893.

Bolnest, Edward. *Medicina Instaurata, or A Brief Account of the true Grounds and Principles of the Art of Physick . . . Also an Epistolary Discourse upon the whole by the Author of Medela Medicinae.* London: John Starkey, 1665.

Bordeu, Théophile de. *Oeuvres complètes de Bordeu, Précédées d'une Notice sur la vie et sur ses Ouvrages par le Chevalier Richerand.* 2 vols. Paris: Caille et Ravier Libraires, 1818.

Borelli, Giovanni Alfonso. *On the Movement of Animals.* Translated by Paul Maquet. Berlin, Heidelberg et al.: Springer-Verlag, 1989.

Borrichius, Olaus. *De ortu et progressu chemiae dissertatio* in the *Bibliotheca chemica curiosa,* Edited by Jean Jacques Manget. 2 vols. Geneva: Chouet, G. De Tournes, et al., 1702., vol. 1, pp. 1–37.

Bostok, Rychard [R.B. Esquire]. *The difference betwene the auncient Phisicke first taught by the godly forefathers, consisting in vnitie peace and concord: and the latter Phisicke proceeding from Idolaters, Ethnickes, and Heathen: as Gallen, and such other consisting in dualitie, discorde, and contrarietie . . .* London: Robert Walley, 1585.

Boulton, Richard, of Brazen-Nose College in Oxford. *An Examination of Mr. Colbatch His Books . . .* London: A. & J. Churchill, 1698.

Boyle, Robert. *The Works of the Honourable Robert Boyle.* 6 vols. London: J. and F. Rivington et al., 1772. Includes Thoms Birch, "The Life of the Honourable Robert Boyle." vol. 1, pp. vi–ccxviii.

Boyle, Robert. *The Sceptical Chymist or Chymico-Physical Doubts and Paradoxes, Touching the Spagyrist's Principles Commonly Call'd Hypostatical: As they are wont to be Propos'd and Defended by the Generality of Alchymists. Whereunto is Praemis'd Part of another Discourse relating to the same Subject.* London: J. Cadwell for J. Crooke, 1661; London: Dawson Reprint, 1965.

Cabriada, Dr. Juan de. *De Los Tiempos y Experiencias el Mejor Remedio al mal. Por la Nova-Antigua Medicina Carta Philosophica Medica Chymica.* Madrid: Lucas Antonio de Bedmar y Baldwin, 1687.

Castle, George, Dr. of Physick, lately Fellow of All-Souls Colledge in Oxon. *The Chemical Galenist: A Treatise Wherein The Practice of the Ancients is reconcil'd to the new Discoveries in the Theory of Physick* . . . London: Sarah Griffin for Henry Twyford, 1667.

Charleton, Walter. *Physiologia Epicuro-Gassendo-Charltoniana* . . ., Introduction and indices by Robert Hugh Kargon. New York and London: Johnson Reprint, 1966.

Clifton, Francis. *The State of Physick, Ancient and Modern, Briefly Consider'd: With a Plan for the Improvement of it.* London: W. Bowyer for John Nourse, 1732.

Coçar, Llorenç. *Dialogus veros medicinae fontes indicans in El "Dialogus" (1589) del Paracelsista Llorenç Coçar y la Cátedra de Medicamentos Quimicos de la Universidad de Valencia (1591).* Introduction by J. M. López Piñero. Valencia: Cátedra e Instituto de Historia de la Medicina, 1977.

Colbatch, John. *A Collection of Tracts Chirurgical and Medical* . . ., 2nd Corrected Edition. London: Dan. Brown, 1704. Includes *A Physico-Medical Essay Concerning Alkaly and Acid, . . . Wherin is endeavoured to be prov'd, that Acids are not (as is generally and erroneously suppos'd) the Cause of all or most Distempers, but that Alkalis are* and *Some Farther Considerations Concerning Alkaly and Acid.*

Conringius, Hermann. *De Hermetica Medicina Libri Duo* . . . 2nd edition, Helmstedt: typis & sumptibus H. Müller, 1669.

Corral, Dr. Don Antonio Alvarez de. *Hippocrates Vindicato y Reflexiones Medicas, sobre el Hippocrates Defendido.* Madrid: Por la Viuda de Juan Garcia Infançon, 1713.

Cotta, John. *A Short Discoverie of the Unobserved Dangers of severall sorts of ignorant and unconsiderate Practisers of Physicke in England.* London: (R. Field) for W. Jones and R. Boyle, 1612.

Coward, William. M.D. of the Colledge of Physicians, London. *Alkali Vindicatum: Or The Acid Opiniator not Guilty of Truth. Being an Impartial Enquiry into the Falacious Reasons and Erroneous Philosophy of a late Physico-Medical Essay Touching Alcali & Acids.* London: Tim. Childe, 1698.

Coward, William. M.D, & Collegii Merton Oxon. Socio. *De Fermento Volatili Nutritio Conjectura Rationalis. Quà ostenditur Spiritum Volatilem Oleosum é Sanguine suffusum, Esse Verum a Genuinum Concoctionis ac Nutritionis Instrumentum.* London: J.H. pro T. Bennet, 1705.

Cowper, Ashley. *The Progress of Physick, by Timothy Scribble, Esq.* London: J. Stagg for J. Roberts, 1743.

Crollius, Oswald. *Basilica Chymica.* Frankfurt: Apud Claudium Marnium & heredes Joannis Aubrii, 1609.

Descartes, René. *Discourse on Method and Other Writings.* Translated with an introduction by F. E. Sutcliffe. Baltimore: Penguin Books, 1968.

Duchesne, Joseph (Joseph Quercetanus). *A Breefe Aunswere . . . to the exposition of Iacobus Aubertus Vindonis, concerning the original, and causes of Metalles. Set foorth against Chimistes. Another exquisite and plaine Treatise of the same Josephus, concerning the Spagericall preparations, and use of minerall, animall, and vegitable secretes, not heertofore knowne of many.* Translated by John Hester, Practitioner in the Spagericall Arte. London, n.p., 1591.

Duchesne, Joseph (Joseph Quercetanus). *The Practise of Chymicall, and Hermeticall Physicke, for the preseruation of health.* Translated by Thomas Tymme. London: Thomas Creede, 1605.

Du Hamel, Jean-Baptiste. *De Consensu Veteris et Novae Philosophiae Libri Duo.* Paris: Apud Carolum Savreux, 1663.

Duncan, Daniel. *La Chymie Naturelle ou L'Explication Chymique et Mechanique de la Nourriture de l'Animal.* Paris: Laurent d'Houry, 1683.

Duncan, Daniel. *Seconde et Troisieme Partie de la Chymie Naturelle, ou L'Explication Chymique et Mechanique De l'Evacuation particuliere aux Femmes, & de la Generation.* Paris: printed at Montaubon: Laurent d'Houry & Daniel Horthemels, 1687.

Emes, Thomas [T.E.], Chirurgo-Medicus. *A Dialogue Between Alkali and Acid: Containing Divers Philosophical and Medicinal Considerations wherein A late Pretended New Hypothesis, asserting Alkali the Cause, and Acid the Cure of all Diseases, is proved Groundless and Dangerous . . .* London: R. Cumberland, 1698.

Emes, Thomas. *A Letter to a Gentleman Concerning Alkali, Being an Answer to a late Piece Intituled, A Letter to a Physician concerning Acid and Alkali.* London: Thomas Speed, 1700.

Fernel, Jean. *Vniversa Medicina, cum notis, observationibus, et remedijs secretis.* Traiecti ad Rhenum [Utrecht]: Typis Gisbert à Zijll, & Theodori ab Ackersdyck, 1656.

Fontaine, Gabriel. *De veritate Hippocraticae Medicinae Firmissimis rationum & experimentorum momentis stabilita, & demonstrata: seu Medicina Antihermetica, In qua Dogmata Medica physiologica, pathologica, & therapeutica, contra Paracelsi, & Hermeticorum placita clarissimé promulgantur: non reiectis penitus Chymicorum inuentis, ad Hippocraticum artem conferentibus. Adiectus est ad calcem generalis index rerum in hoc opere contentarum: necnon introductio ad methodum medendi; atque Apologeticum aduersus Van-helmont, vbi firmissimé demonstratur, quatuor humores Galenistarum non esse fictitios: caetera octauo abhinc folio reconsentur.* Lyon: Philip Bore, Lavr. Arnaud, & Cl. Rigavd, 1657.

Freind, John, D. M. *Praelectiones Chimicae, In quibus omnes fere Operationes Chymicae ad Vera Principia, Et ipsius Naturae Leges rediguntur.* Neapoli: Superiorum permissu, 1722. Printed with Hadrian Mynsicht, *Thesaurus et Armamentarium Medico-Chimicum.*

Freind, John, M.D. *The History of Physick; From the Time of Galen, to the beginning of the Sixteenth Century. Chiefly with Regard to Practice. In a Discourse written to Doctor Mead.* 4th edition, 2 volumes. London: J. Walthoe, 1744, 1750. Reprinted New York: AMS Press, 1973.

Gastaldy, Johannes Baptista. *Quaestio Medica, Proposita ab. Ill. Ac nob. D.D. Joanne-Baprista Gastaldy, An Alimentorum Coctio seu Digestio, à Fermentatione vel à Tritura fiat.* Avenione: Joannem Delorme, 1713.

Gaussen, Isaac. *Réflexions sur l'Application de la Chimie a la Médecine. Présentées à l'École de Médecine de Montpellier, le quatre Thermidor, au six de la République.* Montpellier: Jean Martel, VI (1798).

Glauber, Johann Rudolph. *The Works . . . Containing, Great Variety of Choice Secrets in Medicine and Alchymy . . .* Translated by Christopher Packe. London: Thomas Milbourne, 1689.

Graaf, Regnier de. *De succo Pancreatico: Or a Physical and Anatomical Treatise of the Nature and Office of the Pancreatick Juice . . .* Translated by Christopher Packe, Med. Lond. London: N. Brook, 1676.

Grembs, Franciscus Oswaldus. *Arbor Integra et Ruinosa Hominis, Id est Tractatus Medicus Theorica Practicus . . .* Frankfort: Joh. Georg. Spörlin, 1657.

Guisson, Pierre. M.D. *Epistolica Dissertatio de Anonymo Libello (Circa Abbraeviatum Verae Medicinae Genus) ubi potissimum eventilatur Principiorum Chymicorum Hypothesis* in Pierre Potier. *Opera omnia practica et chymica . . .* Venice: Ex Typographia Balleoniana, 1741, pp. 571–580.

Harris, Thomas. *A Treatise on the Force and Energy of Crude Mercury. Proving the usefulness and innocency of its internal application, but a great variety of experiments and histories of cases, acute and chronick.* London: E. Symon, 1734.

Hecquet, Philippe. *De la Digestion et des Maladies de l'Estomac, suivant le Systeme de la Trituration & du Broyement, sans l'aide de levans ou de la fermentation, don't on fait voir l'impossibilité en santé & en maladie.* Paris: François Fournier, 1712.

Hecquet, Philippe. *De la Digestion et des Maladies de l'Estomac; Suivant le systeme de la Trituration & du Broyement sans l'aide des Levains ou de la Fermentation, don't on fait voir l'imposibilité en santé & en maladie.* Nouvelle ed., revised, corrected and augmented by the author. 2 vols. Paris: Guillaume Cavalier, 1730.

Hecquet, Philippe. *Le Brigandage de la Chirurgie . . . Premier Partie* [and] *Le Brigandage de la Pharmacie . . . Seconde Partie.* Utrecht: Chez les Soeurs de Corneille-Guillaume le Fevre, 1738.

Hecquet, Philippe. *Le Brigandage de la Medecine, dans la maniere de traiter les petites véroles & les plus grandes maladies par l'Emetique, la Saignée, du pié & le Kermés mineral . . .* 3 vols. Utrecht: Corneille Guillaume le Febure, 1749 (1st edition, 1733).

Helmont, Jean Baptiste van. *Ortus medicinae, Id est, initia physicae inaudita. Progressus medicinae novus, in morborum ultionem, ad vitam longam.* Amsterdam: Ludovicus Elsevir, 1648; reprinted Brussels: Culture et Civilisation, 1966. The *Ortus* was published in 1648 with a companion volume, the second edition of the *Opuscula medica inaudita.* Amsterdam: Ludovicus Elsevir, 1648; reprinted Brussels: Culture et Civilisation, 1966.

Helmont, Jean Baptiste van. *Ternary of Paradoxes. The Magnetick Cure of Wounds. Nativity of Tartar in Wine. Image of God in Man.* Translated and ampliated by Walter Charleton. London: James Flesher for William Lee, 1650.

Helmont, Franciscus Mercurius van. *The Paradoxal Discourses of F.M. Van Helmont, Concerning the Macrocosm and the Microcosm, or the Greater and Lesser World And their Union. Set down in Writing by J. B. and now Published.* London: J. C. and Freeman Collins for Robert Kettelwel, 1685.

Hippocrates. *Regimen,* Translated by W. H. S. Jones. *The Loeb Classical Library.* 4 vols. London: Wm. Heineman, 1931.

The History of the Works of the Learned: or An Impartial Account of Books Lately Printed in All Parts of Europe, With a Particular Relation of the State of Learning in Each Country. 12 vols. London: H. Rhodes, J. Harris, T. Bennet, A. Bell, D. Midwinter, and T. Leigh, 1699–1712.

Hoffmann, Friedrich. *Opera omnia physico-medica.* 6 vols. Geneva: de Tournes, 1748.

Hoffmann, Friedrich. *Observationvm Physico-Chymicarvm Selectiorvm Libri III. In qvibvs mvlta cvriosvm experimenta et lectissimae exhibentvr, ad solidam et rationalem chymiam stabiliendam praemissi.* Halle: Prostat in Officina Libraria Rengariana, 1722.

Hoffmann, Friedrich. *Fundamenta medicinae.* Translated and introduced by Lester S. King. London: Macdonald and New York: American Elsevier, 1971.

Irvine, Christopher [C. De Iryngio], Chirurgo-Medecine in the Army. *Medicina Magnetica: or, The rare and wonderful art of Curing by Sympathy: Laid open in Aphorisms: Proved in Conclusions; And digested into an easy Method drawn from both; Wherein the Connexion of the Causes and Effects of these strange Operations, are more fully discovered than heretofore, All cleared and confirmed, by pithy Reasons, true Experiments, and pleasant Relations. Preserved and published, As a Master-piece in this Skill.* No place or publisher. Epistle dedicated Edinburgh 3 June 1656.

Johnson, William, Chymist to the Kings College of Physicians in London. *Αγυρτο - Μαστιξ, or some Brief Animadversions Upon two late Treatises; One of Master George Thomsons, Entituled, Galeno-Pale; The other of Master Thomas O'Dowdes Called The Poor Mans Physician; With a short Appendix, relating to the Company of Apothecaries.* London: T. Mabb for Henry Browne, 1665.

Journal des Sçavans. From 1666, in various editions. Paris and Amsterdam.

Juanini, Jean Baptiste. *Dissertation Physique, ou l'on Montre les mouvements de la fermentation: Les effets des matieres nitreuses dans les corps sublunaires, & les causes que alterent la pureté de l'air de Madrid,* trans. from Spanish to French by Jean-Joseph Courtial. Toulouse: D. Desclasson, 1685.

Juncker, Johann. *Conspectus Chemiae: Élémens de Chymie, suivant les Principes de Becker & de Stahl.* Translated from the Latin of the second edition of M. Juncker with notes by M. De Machy, Apothecaire Gagnant-Maîtrise de l'Hotel-Dieu de Paris. 6 vols. Paris: Siméon-Prosper Hardy, 1757.

Le Clerc, Daniel. *The History of Physick, Or An Account of the Rise and Progress of the Art, And The Several Discoveries therein from Age to Age.* Translated by Dr. Drake and Dr. Boden with Additional Notes. London: D. Brown, A. Rope and J. Leigh, 1699.

Le Clerc, Daniel. *Histoire de la Médecine, Où l'on voit l'Origine & les Progrès de cet Art, de Siècle en Siècle, les Sectes, qui s'y sont formées; les noms des Médecins, leurs découvertes, leurs opinions, & les circonstances les plus remarquables de leur vie.* Nouvelle Edition, Amsterdam: Aux Depens de la Compagnie, 1723. I have also used the 1729 edition with the same pagination.

Lefèvre, Nicholas. *A Complete Body of Chymistry.* Translated by P.D.C. 2 Parts. London: O. Pulleyn, 1670.

Leigh, Charles, M.D. *A Reply to John Colbatch, upon his late Piece, concerning the curing the biting of a viper by acids.* London: Thomas Bennet, 1698.

Mersenne, Marin. *Correspondance du P. Marin Mersenne, Religieux Minime.* Pub. Mme Paul Tannery, ed. Cornelis de Waard with the collaboration of

René Pintard. 12 vols. (1617–1643). Paris: P.U.F. Centre National de la Recherche Scientifique, 1932–1971.

Mongin, Mons. J. *Le Chimiste Physicien* . . . Paris: Laurent d'Houry, 1704.

Morgan, T., M.D. *The Mechanical Practice of Physick: In which The Specifick Method is Examn'd and Exploded; and the Bellinian Hypothesis of Animal Secretion and Muscular Motion, Consider'd and Refuted, With some Occasional Remarks and Scholia on Dr. Lobb's Treatise on the Small Pox, Dr. Robinson on the Animal Oeconomy, and Professor Boerhaave's Account of the Animal Spirits and Muscular Motion.* London: T. Woodward, 1735.

Nedham, Marchamont. *Medela Medicinae, A Plea for the Free Profession, and a Renovation of the Art of Physick, Out of the Noblest and most Authentick Writers. Shewing*

The Publick Advantage of its Liberty.

The Disadvantage that comes to the Publick by any Sort of Physicians, imposing upon The Studies and Practise of them.

The Alteration of Diseases from their Old State and Condition.

The Causes of that Alteration.

The Insufficiency and Uselessness of meer Scholastick Methods and Medicines, with a Necessity of new.

Tending to the Rescue of Mankind from the Tyranny of Diseases; and of Physicians themselves, Pedantism of old Authors and present Dictators. London: Richard Lownds, 1665.

Nicholas Cusanus. *Unity and Reform. Selected Writings of Nicholas de Cusa.* Edited by John P. Dolan. South Bend, Indiana: University of Notre Dame Press, 1962.

Palacios, Felix. *La Pharmacopea Trivnfante de Los Calvmnias y imposturas, que en el Hipocrates Defendido ha publicado el Doctor Don Miguel Boix, Medico honorario, y de la Regia Sociedad Medica de Sevilla.* Madrid: Francisco Martinez Abad, 1713.

Paracelsus. *Sämtliche Werke.* Edited by Karl Sudhoff and Wilhelm Matthiessen. 15 vols. Munich and Berlin: R. Oldenbourg (vols. 6–9: O. W. Barth), 1922–1933.

Paracelsus, *Volumen medicinae paramirum.* Translated and with a preface by Kurt F. Leidecker. *Supplement to Bulletin of the History of Medicine,* no. 11. Baltimore: The Johns Hopkins Press, 1949.

Paracelsus, *Selected Writings.* Edited with an introduction by Jolande Jacobi, Translated by Norbert Guterman. New York: Pantheon Books. Bollingen Series XXVIII, 1951.

Partlicus (Partliz) Von Spitzberg, Simeon. *A New Method of Physick: Or A Short View of Paracelsus and Galen's Practice; In 3 Treatises. I. Opening the Nature of Physick and Alchymy. II. Shewing what things are Requisite to a Physitian and Alchymy. III. Containing a Harmonical Systeme of Physick.* Translated by Nicholas Culpeper, Gent. Student in Physick and Astrologie. London: Peter Cole, 1654.

Patin, Gui. *Lettres . . . Nouvelle Édition augmentée de lettres inédites, précédé d'une notice Biographique . . . par J.-H. Reveillé-Parise.* 3 vols. Paris: J.-B. Baillière, 1846.

Penotus, Bernard G. *"Apologeticall Preface"* to Paracelsus, *A hundred and fourteene Experiments and Cures of the famous Physitian Philippus Aureolus Theophrastus Paracelsus . . . Collected by Iohn Hester.* London: Valentine Sims, 1596.

Pharmacopoeia Augustana Reformata Cum ejus Mantissa & appendice. Simul Animadversionibus, Authore Joanne Zwelfer . . . Cui annexa est ejusdem Authoris Pharmacopoeia Regia Ut & Mantissa Spagyrica. Accessêre etiam huic Editione Bini Discursus Apologetici contra. Otth. Tachenium & Francisc. Verny. Dordrecht: Apud Vincentium Carmax, 1672.

Pharmacopoeia Londinensis. 2nd edition. London: Iohn Marriott, 1618.

Pharmacopoeia Londinensis of 1618 Reproduced in Facsimile with a Historical Introduction by George Urdang. Hollister Pharmaceutical Library Number Two. Madison: State Historical Society of Wisconsin, 1944.

Pitcairn, Archibald. *The Whole Works . . . Wherein are Discovered, The True Foundation and Principles of the Art of Physic. With cases and Observations upon most Distempers and Medicines.* Translated by George Sewell and J.T. Desaguliers. 2nd edition. London: J. Pemberton for W. and J. Innys, 1727.

Pitt, Robert. *The Antidote: Or The Preservative of Health and Life, And the Restorative of Physick To Its Sincerity and Perfection . . .* London: John Nutt, 1704.

Poli, Martino. *Il Trionfo degli Acidi vendicati dalle calumnie di molti Moderni . . .* Rome: Giorgio Placho, 1706.

Poterius, Petrus. *Opera Omnia Practica et Chymica, Cum Annotationibus Vtilissimis Pariter ac Curiosis Friderici Hoffmanni.* Venice: Ex Typographia Balleoniana, 1741.

Quevedo, Francisco de. *Sueños y Discursos.* 2 vols. Madrid: Editorial Castalia, 1993.

Raguet, Sieur. *Dissertation sur Diverses Maladies et sur l'Usage de l'Esprit Philosophique; Avec un Recueil des principales guérisons opérées par ce Remede.* Paris: Aux dépens du Sieur Raguet, Auteur de l'Esprit Philosophique, 1758.

Ribera, Francisco Suarez de [F. Soares da Ribeyra]. *Cirurgia Methodica e Chymica Reformada*, traduzida de Castelhano em Portuguez . . . Joseph Gomez Claro. Lisboa Occidental: Na Officina Ferreyrenciana, 1721.

Ribera, Francisco Suarez de. *Canones Particulares de Cirugia, con que se Libertan Muchos Desahuciados, si al Sagrado de sus fuentes se Refugian.* Madrid: Manuel de Moya, 1751.

Ribera, Francisco Suarez de. *Escrutinio Medico ò Medicina Experimentada* [John Crerar Library copy missing title page. Introductory letter dated 1723].

Ribera, Francisco Suarez de. *Medicina Invincible Legal, O Theatro de Fiebres Intermittentes Complicadas.* Madrid: Francisco del Hierro, 1726.

Ribera, Francisco Suarez de. *Illustracion, y Publicacion de los diez y Siete Secretos del Doctor Juan Curvo Semmedo, confirmadas sus virtudes con maravillosas observaciones.* Madrid: Domingo Fernandez de Arroyo, 1732.

Ribera, Francisco Suarez de. *Amenidades de la Magia Chyrurgica y Medica, Natural.* Madrid: Fernandez de Arroyo, 1736.

Ribera, Francisco Suarez de. *Anatomica Chymica, Inviolable, y Memorable.* Madrid: Manuel de Moya, 1743.

Royal Society of London. *Philosophical Transactions of the Royal Society of London.* From 1665.

Royal Society of London. *The Philosophical Transactions of the Royal Society of London From Their Commencement, in 1665, to the Year 1800; Abridged, with Notes and Biographic Illustrations by Charles Hutton, George Shaw, and Richard Pearson.* 18 vols. London: C. and R. Baldwin, 1809.

Sanctiado, Diego de, Destilador de su Magistad, Vezino de Sevilla. *Arte Separatoria y Modo de Apartar Todos los Licores, que se sacan por via de Distilacion: para que las Medicinas obren conmayor virtud, y prestiza.* Seville: Francisco Perez, 1598.

Senac, Jean Baptiste. *Nouveau Cours de Chymie Suivant les Principes de Newton & de Sthall* (Paris: Jacques Vincent, 1737).

Sennert, Daniel. *De chymicorum cum Aristotelicis et Galenicis consensu ac dissensu liber l. Controversias plurimas tam philosophis quam medicis cognitu utiles continens.* Third edition. Paris: Apud Societatam, 1633.

Sennert, Daniel, Culpeper, Nicholas, and Cole, Abdiah. *Chymistry Made Easie and Useful. Or, The Agreement and Disagreement of the Chymists and Galenists.* London: Peter Cole, 1662.

Servetus, Michael. *A Complete Account of Syrups Carefully Refined According to the Judgment of Galen . . . in A Translation of his Geographical, Medical and Astrological Writings with Introduction and Notes.* Translated by Charles Donald O'Malley. Philadelphia: American Philosophical Society, 1953.

Severinus, Petrus. *Idea medicinae philosophicae.* Third edition. Hagae Comitis: Adrian Vlacq, 1660.

Sherley, Thomas. *A Philosophical Essay: Declaring the Probable Causes, whence Stones are produced in the Greater World. From which occasion is taken to search into the Origin of all Bodies, discovering them to proceed from Water, and Seeds.* London: William Cademan, 1672.

Sprackling, Robert. Doctor of Physick, Coll. Lond. *Medela Ignorantiae or, A Just and plain Vindication of Hippocrates and Galen from The groundless Imputations of M.N. wherein The whole Substance of his Illiterate Plea, intituled Medela Medicinae is occasionally considered.* London: W. G. for Robert Crofts, 1665.

Stahl, Georg Ernst. *Oeuvres Médico-Philosophique et Pratiques de G.-E. Stahl.* Translated and commentary by T. Blondin. Second edition. Paris/Montpellier/Strasbourg: J. B. Ballière et Fils/Pitrat/Treuttel et Wurtz, 1864. Volumes two through six (of a planned ten) of the Blondin translation were published.

Stahl, Georg Ernst. *Paraenesis, ad Aliena a Medica Doctrina Arcendvm; Quoniam non Solvm Medicam Doctrinam Morentvr, Tvrbent, Fallant, sed Aliena ab ipsa Veritate ey vsv Practico, Animis Obiiciant.* Halle: Orphanotrophei, 1706.

Stahl, Georg Ernst. *Disqvisitio de Mechanismi et Organismi Diversitate* . . . Halle: Sumptibus & Impensis Orphanotrophei, 1706.

Stahl, Georg Ernst. *Vera Diversitate Corporis Mixti et Vivi, et Vtrivsque Peculiarum Conditionvm atqve Proprietatvm Necessaria Discretione Demonstratio.* Halle Magdeburg: Sumptibus Litterisqve Orphanotrophei, 1707.

Stahl, Georg Ernst. *Negotiosum Otiosum, seu σκια μαχια adversus positiones aliquas fundamentales theoriae verae medicae.* Halle: Litteris et Impensis Orphanotrophei, 1720.

Stahl, Georg Ernst. *Philosophical Principles of Universal Chemistry: or The Foundation of a scientifical Manner of Inquiring into and Preparing The Natural and Artificial Bodies for the Uses of Life: Both in the smaller Way of Experiment, and the larger Way of Business. Design'd as a General Introduction To the Knowledge and Practice of Artificial Philosophy: Or Genuine Chemistry in all its Branches. Drawn from the Collegium Jenense of Dr. George Ernest Stahl. By Peter Shaw, M.D.* London: John Osborn and Thomas Longman, 1730.

Stahl, Georg Ernst. *Fundamenta Chymiae Dogmaticae et Experimentalis* . . . Nuremberg: Imbensis B. Guolfg. Maur. Endteri, 1746.

Struve, Ernest Gotthold. *Paradoxum Chymicvm sive Igne i.e. Operationes et Experimenta Physico-Chymico-Pharmacevtica, ipsaqve Medicamenta Chymica Ignis Operari Solita Sive Igne.* Jena: Ernest. Clavd. Bailliae, c. 1715.

Stubbe, Henry. *The Lord Bacons Relation to the Sweating-Sickness Examined, in a Reply to George Thomson, Pretender to Physick and Chymistry. Together with a Defence of Phlebotomy in General, and also particularly In the Plague, Small-pox, Scurvy and Pleurisy. In opposition to the same Author, and the Author of Medela Medicinae, Doctor Whitaker, and Doctor Sydenham* . . . London: Phil. Brigs, 1671.

[Swift, Jonathan], *A Tale of a Tub. Written for the Universal Improvement of Mankind. To which is added, An Account of a Battel Between the Ancient and Modern Books in St. James Library.* The Fourth Edition Corrected. London: Printed for John Nutt, 1725.

Sydenham, Thomas, M.D. *The Works of Dr. Thomas Sydenham, Newly made English from the Originals: Wherein the History of Acute and Chronic Diseases, and the Safest and most Effectual Methods of treating them, are faithfully, clearly, and accurately delivered. To which are added, Explanatory and Practical Notes, From the best medicinal Writers.* The Second Edition, Revised, Corrected, and Enlarged with several Additional Notes by John Swan, M.D. London: Edward Cave, 1749.

Sylvius, Franciscus de la Boë. *Opera medica.* Amsterdam: Daniel Elsevier and Abraham Wolfgang, 1680.

Sylvius, Franciscus de la Boë. *A New Idea of the Practice of Physic; . . . The First Book; Of the diseases either Constitutions, producing, or following the Natural Functions of Man not in Health. Wherein is containd, beside a new Method in general, a Vindication of the Spleen and Mother from Fits attributed to them. As also a new Discovery of Intermitting Fevers, the Yellow Jaundice, and other Diseases never before discoverd. All cleared by Anatomical Experiments, and Chymical Demonstrations; As also by their Cures. Whereto is prefixed a Preface written by Dr. Mar. Nedham. Translated by Richard Gower, formerly Student under the Author.* London: Brabazon Aylmer, 1675.

Tachenius, Otto. *Hippocrates Chymicus, Discovering the Ancient foundations of the late Viperine Salt with his Clavis there unto annexed.* Translated by I.W. London: Nath. Crouch, 1677.

Tachenius, Otto. *His Clavis to the Ancient Hippocratical Physick or Medicine; Made by Manual Experience in the very Fountains of Nature. Whereby, Through Fire and Water, in a Method unheard of before, the Occult Mysteries of Nature and Art are Unlocked and clearly Explained by a Compendious way of Operation.* London: Printed by Tho. James and are to be sold by Nath. Crouch, 1677.

Theatrum Sympatheticum auctum, exhibens varios authores de pulvere sympathetico . . . Nuremberg: Joh. Andreas Endter, 1662.

J.T. [James Thompson] Esq. *Helmont Disguised: or The Vulgar Errours of Impericall and unskillfull Practisers of Physick confuted. More especially, As they concern the cures of Feavers, Stone, Plague, And other Diseases. In a Dialogue between Philiatros and Pyrosophilus. In which the chief Rarities of Physick is admirably discoursed of.* London: E. Alsop for N. Brook, and W. Leybournse, 1657.

Tixedas, Christoval. *Verdad Defendida, Respuesta de Fileatro, a Carta Medico-Chymica, que contra los Medicos de la Iunta, de la Corte, y contra todos los Galenicos, le escriviò el Doctor Medico-Chymico D. Iuan Cabriada.* Barcelona: En casa de Antonio Ferrer y Balthasar Ferrer, 1688.

Tuthill, Francis. *A Vindication of Some Objections lately raised against Dr. John Colbatch his Hypothesis.* 1698.

Twysden, John. M.D. Fellow of Kings College of Physicians in London. *Medicina Veterum vindicata: Or an Answer To a Book, entituled Medela Medicinae* . . . London: J.G. for John Crook, 1666.

Uwins, David. *Modern Medicine, Containing a Brief Exposition of the Principal Discoveries and Doctrines that Have Occasioned the Recent Advancement of Medical Philosophy, with Strictures on the Present State of Medical Practice* . . . London: Samuel Tipper, 1808.

Venel, Gabriel François. "Chymie" in *Encyclopédie, ou Dictionnaire Raissonné des Sciences, des Arts et des Métiers.* Edited by M. Diderot and M. D'Alembert. Vol. 3. Paris: Briasson, David, Le Breton and Durand, 1753.

Vieussens, Raymond. Medecin de la Faculté de Montpellier. *Deux dissertations . . . Le Premiere Touchant l'extraction du Sel acide du Sang, La Second Sur la proportion de quantité de ses principes sensibles.* Montpellier: Honoré Pech, 1698.

Vieussens, Raymond. Medecin ordinaire du Roi, de la Académie des Sciences de Paris, & de la Societé Royale de Londres. *Traité Nouveau de la Structure et des Causes du Mouvement Natural du Coeur.* 2 vols. Toulouse: Jean Guillemette, 1715.

Willis, Thomas. *Dr. Willis's Practice of Physick, Being the whole Works of that Renowned and Famous Physician* . . . London: T. Dring, C. Harper and J. Leigh, 1681; also 1684 with reset type.

Willis, Thomas. *Opera Omnia.* Venice: Sumptibus Ruïnetti, & Storti, 1708.

S.W. *An Examination of a late Treatise of the Gout. Wherein John Colbatch's demonstration are briefly refuted.* 1698.

Zapata, Diego Mateo. *Verdadera Apologia en Defensa de la Medicina Racional Philosophica* . . . Madrid: Antonio de Zafra, 1690.

B. SECONDARY SOURCES

Alves Porto, Paulo. *Van Helmont e o Conceito de Gás: Química e Medicina no sécolo XVII.* São Paulo: Educ, 1995.

Anon. "La Historia de la Ciencia en España como Realidad Marginal en su Organización y Contexto Social." *Anthropos.* No. 20 (1982), 2–15.

Balaguer Perigüell, Emilio. *La Introducción del Modelo Fisico-Matemático en la Medicina Moderna: Análisis de la obra de G.A. Borelli (1608–1679) De motu animalium, Cuadernos Hispánicos de Historia de la Medicina y de la Ciencia, XIV,* Serie A (Monografías). Valencia-Granada, 1979.

Baldwin, Martha. "Expanding the Therapeutic Canon: Learned Medicine Listens to Folk Medicine." Unpublished paper presented at Duke University, 18 April 1998.

Baumann, E. D. *François de la Boë Sylvius.* Leiden: E. J. Brill, 1949.

Boas, Marie. "An Early Version of Boyle's *Sceptical Chymist.*" *Isis, 45* (1954), 153–68.

Boas, Marie. "Acid and Alkali in Seventeenth Century Chemistry." *Archives Internationales d'Historire des Sciences, 9* (1956), 13–28.

Boas, Marie. *Robert Boyle and Seventeenth-Century Chemistry.* Cambridge: Cambridge University Press, 1958.

Boas Hall, Marie. "Otto Tachenius." In *Dictionary of Scientific Biography,* vol. 13, pp. 234–35. New York: Scribner's, 1971.

Brockliss, Laurence. "The Medico-Religious Universe of an Early 18th Century Parisian Doctor: The Case of Philippe Hecquet." In *The Medical Revolution of the 17th Century.* Edited by Roger French and Andrew Wear. Cambridge: Cambridge University Press, 1989, pp. 191–221.

Brockliss, Laurence and Jones, Colin. *The Medical World of Early Modern France.* Oxford: Clarendon Press, 1997.

Broeckx, C. *Commentaire de J.B. van Helmont sur le premier livre du Régime d'Hippocrate: Peri diaites.* Antwerp, 1849.

Broeckx, C. *Interrogatoires de J.B. van Helmont sur le magnétisme animal.* Anvers: Buschmann, 1856.

Brown, Theodore M. "Lorenzo Bellini." In *Dictionary of Scientific Biography,* vol. 1, pp. 592–94. New York: Scribner's, 1970.

Brown, Theodore M. "Archibald Pitcairne." In *Dictionary of Scientific Biography,* vol. 11, pp. 1–3. New York: Scribner's, 1975.

Brown, Theodore M. "Physiology and Mechanical Philosophy in Mid-Seventeenth Century England." *Bulletin of the History of Medicine, 51* (1977), 27–54.

Bujanda, J. M. *Index de L'Inquisition Espagnole (1583, 1584)*. Quebec: Centre d'Études de la Renaissance: Éditions de l'Université de Sherbrooke/ Libraire Droz, 1993.

Burton, William. *An Account of the Life and Writings of Herman Boerhaave . . . In Two Parts, with an Appendix*. London: Henry Lintot, 1743.

Carter, Richard B. *Descartes' Medical Philosophy: The Organic Solution to the Mind-Body Problem*. Baltimore and London: The Johns Hopkins Press, 1983.

Christie, J. R. R. "Historiography of Chemistry in the Eighteenth Century: Hermann Boerhaave and William Cullen." *Ambix*, *41* (1994), 4–19.

Clark, Sir George. *A History of the Royal College of Physicians of London*. 3 vols. Oxford: Clarendon Press, 1954, 1966, 1972.

Clericuzio, Antonio. "A Redefinition of Boyle's Chemistry and Corpuscular Philosophy." *Annals of Science*, *47* (1990), 561–69.

Clericuzio, Antonio. "From van Helmont to Boyle, A Study of the Transmission of Helmontian Chemical and Medical Theories in Seventeenth Century England." *British Journal of the History of Science*, *26* (1993), 303–34.

Clericuzio, Antonio. "Carneades and the Chemists: a Study of the *Sceptical Chymist* and its Impact on Seventeenth Century Chemistry." In *Robert Boyle Reconsidered*, Edited by Michael Hunter. Cambridge: Cambridge University Press, 1994, pp. 79–90.

Cook, Harold J. *The Decline of the Old Medical Regime in Stuart London*. Ithaca and London: Cornell University Press, 1986.

Cook, Harold J. "The Society of Chemical Physicians. The New Philosophy and the Restoration Court." *Bulletin of the History of Medicine*, *61* (1987), 61–77.

Cook, Harold J. "The Rose Case Reconsidered: Physicians, Apothecaries and the Law in Augustan England." *Journal of the History of Medicine and Allied Sciences*, *45* (1990), 527–55.

Cook, Harold J. "Sir John Colbatch and Augustan Medicine: Experimentalism, Character and Entrepeneuralism." *Annals of Science*, *47* (1990), 475–505.

Costabel, Pierre. "Jean-Baptiste Du Hamel." In *Dictionary of Scientific Biography*, vol. 4, pp. 221–22. New York: Scribner's, 1971.

Davis, Audrey B. *Circulation Physiology and Medical Chemistry in England 1650–1680*. Lawrence, Kansas: Coronado Press, 1973.

Debus, Allen G. "An Elizabethan History of Medical Chemistry." *Annals of Science*, *18* (1962, published 1964), 1–29.

Debus, Allen G. "The Paracelsian Aerial Niter." *Isis, 55* (1964), 43–61.

Debus, Allen G. "The Significance of the History of Early Chemistry." *Cahiers d'Histoire Mondiale, 23* (1965), 39–58.

Debus, Allen G. *The English Paracelsians.* London: Oldbourne, 1965.

Debus, Allen G. "Fire Analysis and the Elements in the Sixteenth and Seventeenth Centuries," *Annals of Science, 23* (1967), 127–47.

Debus, Allen G. "Mathematics and Nature in the Chemical Texts of the Renaissance." *Ambix, 15* (1968), 1–28, 211.

Debus, Allen G. "Edward Jorden and the Fermentation of the Metals: An Iatrochemical Study of Terrestrial Phenomena." In *Toward a History of Geology: Proceedings of the New Hampshire Interdisciplinary Conference on the History of Geology. September 7–12, 1967.* Edited by Cecil J. Schneer. Cambridge, Mass.: M.I.T. Press, 1969. Pp. 100–21.

Debus, Allen G. *The Chemical Philosophy: Paracelsian Science and Medicine in the Sixteenth and Seventeenth Centuries.* 2 vols. New York: Science History Publications, 1977.

Debus, Allen G. *Man and Nature in the Renaissance.* Cambridge University Press, 1978.

Debus, Allen G. "Mysticism and the Rise of Modern Science." *Journal of Central Asia, 3* (1980), 46–61.

Debus, Allen G. "Thomas Sherley's 'Philosophical Essay' (1672)." *Ambix, 27* (1980), 124–35.

Debus, Allen G. "Chemistry and the Quest for a Material Spirit of Life in the Seventeenth Century." In *Spiritus: IV Colloquio Internazionale del Lessico Intellettuale Europeo, Roma, 7–9 gennaio 1983.* M. Fattori and M. Bianchi, editors. Roma: Edizioni dell'Ateneo, 1984, pp. 245–63.

Debus, Allen G. "History With a Purpose: The Fate of Paracelsus." *Pharmacy in History, 26* (1984), 83–96.

Debus, Allen G. "Chemistry and the Universities in the Seventeenth Century." *Academiae Analecta: Klasse der Wetenschappen, 48* (1986), 13–33.

Debus, Allen G. "Chemistry and Iatrochemistry in Early Eighteenth Century Portugal: A Spanish Connection." In *Historia e Desenvolvimento da Ciência em Portugal: I Colóquio - até ao Sécolo XX. Lisboa, 15 a 19 Abril de 1985.* Secretary-General of the Colloquium, Prof. Doutor António Vasconcellos Marques. 2 vols. Lisboa: Academia das Ciências de Lisboa, 1986, vol. 2, pp. 1245–62.

Debus, Allen G. *The French Paracelsians: The Chemical Challenge to Medical and Scientific Tradition in Early Modern France.* Cambridge University Press, 1991.

Debus, Allen G. "Paracelsus and the Medical Revolution of the Renaissance: A 500th Anniversary Celebration." In *Paracelsus, Five Hundred Years: Three American Exhibits.* Bethesda, Maryland: Published by the Friends of the National Library of Medicine, Inc., for the Hahnemann University Library, The National Library of Medicine, and the Washington University School of Medicine, St. Louis, 1993, pp. 3–12.

Debus, Allen G. *Paracelso e la Tradizione Paracelsiana.* Napoli: La Città del Sole, 1996.

Debus, Allen G. "Paracelsus and the Delayed Scientific Revolution in Spain: A Legacy of Philip II." In *Reading the Book of Nature: The Other Side of the Scientific Revolution.* Edited by Allen G. Debus and Michael T. Walton. Volume XLI, Sixteenth Century Essays and Studies. Kirksville, Missouri: Thomas Jefferson University Press, 1997, pp. 139–52.

Debus, Allen G. "The Rise and Fall of Chemical Physiology in the Seventeenth Century." *Memórias da Academia das Ciências de Lisboa - Classe de Ciências, 36* (1996/97, published 1998), 35–60.

Debus, Allen G. "Paracelsianism and the Diffusion of the Chemical Philosophy in Early Modern Europe." In Ole Peter Grell, Editor. *Paracelsus: The Man and His Reputation, His Ideas and Their Transformation.* Leiden, Boston, Köln: Brill, 1998, pp. 225–44.

Debus, Allen G. "The Chemical Debates of the 17th Century: The Embattled New Science of the Chemists." In *Anais: V Seminário Nacional da Ciência e da Tecnologia: III Reunião da Rede de Intercâmbios para a Historia e a Epistemologia das Ciências Químicas e Biológicas: 24 a 28 de Julho de 1995 Ouro Preto MG.* São Paulo: Sociedade Brasileira de História de Ciência, 1998, pp. 7–19.

Debus, Allen G. "Chemists, Physicians, and Changing Perspectives on the Scientific Revolution." *Isis, 89* (1998), 66–81.

Dictionary of Scientific Biography. Charles C. Gillispie, Editor-in-Chief. 16 vols. New York: Scribner's, 1970–1980.

Donovan, Arthur, editor. *The Chemical Revolution: Essays in Reinterpretation. Osiris,* Second Series, vol. 4 (1988).

Duchesneau, François. "Leibniz et Stahl: Divergences sur le concept d'organisme." *Stud. Leibnitiana, 27* (1995), 185–212.

Fantini, Bernardino. "Iatrofisica e Iatrochimica: La Fisica, La Chimica ed I Modelli dell'Organizzazione del Vivente." *Medicina nei Secoli: Arte e Scienza (Journal of History of Medicine), 9* (1997), 59–95.

Firth, C. H. "Marchamont Nedham." In *Dictionary of National Biography,* vol. 14. Oxford University Press, 1920–1921, pp. 159–64.

Fisch, Harold. "The Scientist as Priest: A Note on Robert Boyle's Natural Theology." *Isis, 44* (1953), 252–65.

Frank, Robert G., Jr. *Harvey and the Oxford Physiologists: A Study of Scientific Ideas.* Berkeley and Los Angeles: University of California Press, 1980.

Frank, Robert G., Jr. "Thomas Willis." In *Dictionary of Scientific Biography*, vol. 14, pp. 404–9. New York: Scribner's, 1976.

Fulton, John F. *A Bibliography of The Honourable Robert Boyle, Fellow of the Royal Society.* 2nd edition. Oxford: Clarendon Press, 1961.

Garrison, Fielding H. *An Introduction to the History of Medicine with Medical Chronology, Suggestions for Study and Bibliographic Data.* Fourth edition. Philadelphia and London: W.B. Saunders Co., 1961.

Gelbart, Nina Rattner. "The Intellectual Development of Walter Charleton." *Ambix, 18* (1971), 149–68.

Geyer-Kordesch, Johanna. "Chemie und Alchemie: J.J. Becher, G.E. Stahl und J.C. Carl und J.C. Dippel." In *Johann Becher (1635–1682).* Edited by Gotthardt Fruhsorge und Gerhard F. Strasser. Wiesbaden: Otto Harrassewitz, 1993, pp. 127–42.

Gillispie, Charles Coulston. *The Edge of Objectivity.* Princeton University Press, 1960.

Gillispie, Charles Coulston. "The *Encyclopédie* and the Jacobin Philosophy of Science." In *Critical Problems in the History of Science.* Edited by Marshall Clagett. Madison: University of Wisconsin Press, 1962, pp. 255–89.

Godwin, Gordon. "Sir George Castle." In *Dictionary of National Biography*, vol. 3, p. 1183. Oxford University Press, 1920–1921.

Godwin, Gordon. "Thomas Emes." In *Dictionary of National Biography*, vol. 6, pp. 772–73. Oxford University Press, 1920–1921.

Godwin, Joscelyn. *Robert Fludd: Hermetic Philosopher and Surveyor of Two Worlds.* London: Thames and Hudson, 1979.

Gough, J. B. "Lavoisier and the Fulfillment of the Stahlian Revolution." *Osiris, 4* (1988), 15–33.

Granit, Ragnar. "Jean Fernel." In *Dictionary of Scientific Biography*, vol. 4, pp. 584–86. New York: Scribner's, 1971.

Granjel, Luis S. *Francisco Suarez de Rivera: Médico salamantino del siglo XVIII.* Salamanca: University of Salamanca Press: Cuadernos de Historia de la Medicina Española, Monografias IV, 1967.

Grmek, Mirko D. "Giorgio Baglivi." In *Dictionary of Scientific Biography*, vol. 1, pp. 391–92. New York: Scribner's, 1970.

Grmek, Mirko D. "Raymond Vieussens." In *Dictionary of Scientific Biography*, vol. 14, pp. 25–27. New York: Scribner's, 1976.

Grmek, Mirko D. *La première révolution biologique: Reflexions sur la physiologie et la medecine du xviie siècle.* Paris: Editions Payot, 1990.

Grmek, Mirko D. "La vita e l'opera di Giorgio Baglivi, medico raguseo e leccese." In *Il nucleo filosofico della scienza*, Guido Cimino et al., editors. Galatina: Congedo, 1991, pp. 93–111.

Guerrini, Anita. "James Keill, George Cheyne, and Newtonian Philosophy, 1690–1740." *Journal of the History of Biology, 18* (1985), 247–66.

Guerrini, Anita. "The Tory Newtonians: Gregory, Pitcairne, and Their Circle." *Journal of British Studies, 25* (1986).

Guerrini, Anita. "Archibald Pitcairne and Newtonian Medicine." *Medical History, 31* (1987), 70–83.

Gunnoe, Charles D. "Erastus and Paracelsianism: Theological Motifs in Thomas Erastus' Rejection of Paracelsian Natural Philosophy." In *Reading the Book of Nature: The Other Side of the Scientific Revolution.* Allen G. Debus and Michael T. Walton, editors. Kirksville, Missouri: Thomas Jefferson University Press, 1998, pp. 45–66.

Haigh, Elizabeth. *Xavier Bichat and the Medical Theory of the Eighteenth Century. Medical History.* Supplement 4. London: Wellcome Institute for the History of Medicine, 1984.

Hall, A. Rupert. *From Galileo to Newton 1630–1720.* New York and Evanston: Harper & Row, 1963.

Hannaway, Owen. *The Chemists and the Word: The Didactic Origins of Chemistry.* Baltimore: The Johns Hopkins Press, 1975.

Hannaway, Owen. "Laboratory Design and the Aim of Science: Andreas Libavius versus Tycho Brahe." *Isis,* 77 (1986), 585–610.

Harley, David. "Rychard Bostok of Tandridge, Surrey (c. 1530–1605), M.P., Paracelsian Propagandist and Friend of John Dee." *Ambix, 47* (2000), 29–36.

Heineke, Berthold. *Wissenschaft und Mystik bei J.B. van Helmont (1579–1644).* Bern, Berlin, et al.: Peter Lang, 1996.

Hoffmann, Paul. "La controverse entre Leibniz et Stahl sur le nature de l'âme." *Studies on Voltaire and the Eighteenth Century, 199* (1981), 237–41.

Holmes, Frederic Lawrence. *Lavoisier and the Chemistry of Life: An Exploration of Scientific Creativity.* Madison, The University of Wisconsin Press, 1985.

Holmes, Frederic Lawrence. *Eighteenth-Century Chemistry as an Investigative Enterprise.* Berkeley: University of California: Office for History of Science and Technology, 1989.

Howard, Rio. "Guy de la Brosse and the Jardin des Plantes in Paris." In *The Analytic Spirit: Essays in the History of Science in Honor of Henry Guerlac.*

Edited by Harry Woolf. Ithaca: Cornell University Press, 1981, pp. 195–224.

Howe, H. M. "A Root of van Helmont's Tree." *Isis, 56* (1965), 408–19.

Huard, Pierre. "Jean Astruc." In *Dictionary of Scientific Biography*, vol. 1, pp. 322–24. New York: Scribner's, 1970.

Huffman, William H. *Robert Fludd and the End of the Renaissance.* London and New York: Routledge, 1988.

Hunter, Michael. "Alchemy, Magic and Moralism in the Thought of Robert Boyle." *British Journal of the History of Science, 23* (1990), 387–410.

Hunter, Michael, editor. *Robert Boyle Reconsidered* (Cambridge University Press, 1994).

Hunter, Michael. "Boyle versus the Galenists: An Aborted Revolution in Seventeenth-Century Medicine." As yet unpublished paper read in typescipt.

Innes-Smith, R. W. *English Speaking Students at the University of Leiden.* Foreword by John W. Comrie. Edinburgh: Oliver and Boyd, 1932.

Isler, Dr. Hansruedi. *Thomas Willis 1621–1675. Doctor and Scientist.* New York/London: Hafner, 1968. First edition in German, 1965.

Jones, Richard Foster. *Ancients and Moderns: A Study of the Rise of the Scientific Movement in Seventeenth-Century England.* St. Louis: Washington University Press, 1963. First edition, 1936.

Kamen, Henry. *Spain 1469–1714: A Society in Conflict.* London and New York: Longman, 1983.

Kamen, Henry. *Spain in the Later Seventeenth Century 1665–1714.* London and New York: Longman, 1983.

Kaplan, Barbara Beigun. *Divulging of Useful Truths in Physick: the Medical Agenda of Robert Boyle.* Baltimore: The Johns Hopkins Press, 1993.

King, Lester S. "Stahl and Hoffmann: A Study in Eighteenth Century Animism." *Journal of the History of Medicine and Allied Sciences, 19* (1964), 118–130.

King, Lester S. "Boissier de Sauvages and 18th Century Nosology." *Bulletin of the History of Medicine, 40* (1966), 43–51.

King, Lester S. *The Road to Medical Enlightenment 1650–1695.* London: Macdonald/New York: American Elsevier, 1970.

King, Lester S. "Georg Ernst Stahl." In *Dictionary of Scientific Biography*, vol. 12, pp. 599–606. New York: Scribner's, 1975.

King, Lester S. *The Philosophy of Medicine: The Early Eighteenth Century.* Cambridge and London: Harvard University Press, 1978.

Klickstein, Herbert S. "Charles Caldwell and the Controversy in America Over Liebig's *Animal Chemistry.*" *Chymia, 4* (1953), 129–57.

Lemoine, Albert. *Le vitalisme et l'animisme de Stahl.* Paris: Germer Baillière, 1864.

Levine, Joseph M. *Dr. Woodward's Shield: History, Science, and Satire in Augustan England.* Berkeley, Los Angeles, London: University of California Press, 1977.

Levine, Joseph M. *The Battle of the Books: History and Literature in the Augustan Age.* Ithaca and London: Cornell University Press, 1991.

Lindeboom, G. A. *Herman Boerhaave. The Man and His Work.* London: Methuen & Co., Ltd., 1968.

Lindeboom, G. A. "Herman Boerhaave." In *Dictionary of Scientific Biography,* vol. 2, pp. 224–28. New York: Scribner's, 1970.

Lindeboom, G. A. "Franciscus de la Boé Sylvius." In *Dictionary of Scientific Biography,* vol. 13, pp. 222–23. New York: Scribner's, 1976.

López Piñero, José María. *La Introducción de la Ciencia Moderna en España.* Barcelona: Ediciones Ariel, 1969.

López Piñero, José María. "El Primer Sistema Medico Moderno: La Iatroquimica de la Segunda Mitad del Siglo XVII." "I. Posición histórica del movimiento. Silvio y la iatroquímica holandesa y alemana." *Medicina Española, 67* (1972), 164–73; "II. Willis y la iatroquímica inglesa y los países latinos. La crisis del movimiento." *Medicina Española, 67* (1972), 228–37.

López Piñero, José María. *Ciencia y Técnica en la Sociedad Española de los Siglos XVI y XVII.* Barcelona: Editorial Labor, 1979.

López Piñero, José María. *Clasicos Médicos Valencianos del Siglo XVI.* Valencia: Generalitat Valenciana, 1990.

López Piñero, José María, Glick, Thomas F., Navarro Brotóns, Victor, and Portela Marco, Eugenio. *Diccionario histórico de la ciencia moderna en España.* 2 vols. Barcelona: Ediciones Península, 1983.

Marx, Jacques. "Alchimie et Palingénésie." *Isis, 62* (1971), 274–89.

Merck Luengo, José Guillermo. *La Quimiatria en España.* Madrid: Instituto "Arnaldo de Vilanova," 1959.

Metzger, Hélène. *Les doctrines chimiques en France du début XVIIe siécle. Premiere partie.* Paris: P.U.F., 1923.

Metzger, Hélène. *Newton, Stahl, Boerhaave et la doctrine chimique.* Paris: Félix Alcan, 1930.

Mévergnies, Nève de. *Jean-Baptiste van Helmont, Philosophe par le feu.* Paris: Libraire E. Croz, 1935.

Moe, Harold. "When Steno Brought New Esteem to Glands." In *Nicholas Steno 1638–1686: A Reconsideration by Danish Scientists.* Edited by Jacob E. Poulsen and Egill Snorrason. Gentofte, Denmark: Nordisk Insulinlaboratorium, 1986, pp. 51–96.

Moore, Norman. "John Twysden." In *Dictionary of National Biography*, vol. 19, p. 1331. Oxford University Press, 1920–1921.

Moran, Bruce T. *Chemical Pharmacy Enters the University: Johannes Hartmann and the Didactic Care of Chymiatria in the Early Seventeenth Century.* Madison: American Institute of the History of Pharmacy, 1991.

Müller, Ingo W. "Untersuchungen zum Hippokratesverständnis von Otto Tachenius und Michael Ettmüller." *Medizinische Journal, 22* (1987), 327–341.

Multhauf, Robert P. "Medical Chemistry and 'The Paracelsians'." *Bulletin of the History of Medicine, 28* (1954), 101–26.

Multhauf, Robert P. "Johann Zwelfer." In *Dictionary of Scientific Biography*, vol. 14, pp. 639–40. New York: Scribner's, 1976.

Munk, William. *The Roll of the Royal College of Physicians of London* . . . 3 vols., 2nd edition. London: Published by the College, 1878.

Nash, Leonard K. *Plants and the Atmosphere.* Case 5. Harvard Case Histories in Experimental Science, James Bryant Conant, General Editor. Cambridge, Mass.: Harvard University Press, 1952.

Newman, William. "Boyle's Debt to Corpuscular Alchemy." In Michael Hunter, *Robert Boyle Reconsidered.* Cambridge University Press, 1994, pp. 107–17.

Niebyl, P. H. "Galen, Van Helmont and Blood Letting." In *Science, Medicine and Society in the Renaissance. Essays to Honor Walter Pagel.* Edited by Allen G. Debus. 2 vols. New York: Science History Publications/ London: Heinemann, 1972. Vol. 2, pp. 13–23.

Pagel, Walter. "Helmont, Leibniz, Stahl." *Archiv für Geschichte der Medizin, 24* (1931), 19–59.

Pagel, Walter. "Van Helmont's Ideas on Gastric Digestion and Gastric Acid." *Bulletin of the History of Medicine, 30* (1956), 524–36.

Pagel, Walter. *Paracelsus: An Introduction to Philosophical Medicine in the Era of the Renaissance.* Basel: S. Karger, 1958; second revised edition, Basel: S. Karger, 1982.

Pagel, Walter. *William Harvey's Biological Ideas: Selected Aspects and Historical Background.* Basel/New York: S. Karger, 1967.

Pagel, Walter. "Van Helmont's Concept of Disease–To Be or Not To Be? The Influence of Paracelsus." *Bulletin of the History of Medicine, 46* (1972), 419–54.

Pagel, Walter. *Joan Baptista Van Helmont; Reformer of Science and Medicine.* Cambridge University Press, 1981.
Pagel, Walter. *The Smiling Spleen: Paracelsianism in Storm and Stress.* Basel et al.: S. Karger, 1984.
Partington, J. R. *A Short History of Chemistry.* London: Macmillan, 1951.
Partington, J. R. *A History of Chemistry.* Vol. 2. London: Macmillan, 1961; Vol. 3. London: Macmillan, 1962.
Patterson, T. S. "Jean Beguin and his *Tyrocinium Chymicum.*" *Annals of Science, 2* (1937), 243–98.
Pilpoul, Pascal. *La Querelle de l'Antimoine (Essai historique).* Thèse pour l'Doctorat en Médecine. Paris: Librairie Louis Arnette, 1928.
Prieto Aguirre, José F. "La Obra de Boix y Moliner. Historia de una Polemica." *Estudios de Historia de la Medicina Española.* Nueva Series I (1959–1960), no. 6.
Principe, Lawrence M. "The Gold Process: Directions in the Study of Robert Boyle's Alchemy." In Z. R. W. M. von Martels, editor. *Alchemy Revisited: Proceedings of the International Conference on the History of Alchemy at the University of Groningen 17–19 April 1989.* Leiden et al.: E. J. Brill, 1990, pp. 200–6.
Principe, Lawrence M. "Robert Boyle's Dialogue on Transmutation." Unpublished paper presented at the History of Science Meeting in Madison, Wisconsin, November, 1991.
Principe, Lawrence M. "Robert Boyle's Alchemical Pursuits." In *Robert Boyle Reconsidered.* Edited by Michael Hunter. Cambridge University Press, 1994, pp. 91–106.
Principe, Lawrence M. *The Aspiring Adept: Robert Boyle and his Alchemical Quest.* Princeton University Press, 1998.
Rather, L. J. and Frerichs, J. B. "The Leibniz-Stahl Controversy. I. Leibniz' Opening Objections to the *Theoria medica.*" *Clio medica, 3* (1968), 21–40.
Rather, L. J. and Frerichs, J. B. "The Leibniz-Stahl Controversy. II. Stahl's Survey of the Principal Points of Debate." *Clio medica, 5* (1970), 53–67.
Rattansi, P. M. "The Helmontian-Galenist Controversy in Restoration England." *Ambix, 12* (1964), 1–23.
Risse, Guenter B. "Friedrich Hoffmann." In *Dictionary of Scientific Biography,* vol. 6, pp. 458–61. New York: Scribner's, 1972.
Rommelaere, Willem. *Études sur J.B. Van Helmont.* Brussels: Manceaux, 1868.
Sargent, Rose-Mary. *The Diffident Naturalist: Robert Boyle and the Philosophy of Experiment.* Chicago: The University of Chicago Press, 1995.
Secret, François. "Palingenesis, Alchemy and Metempsychosis in Renaissance Medicine." *Ambix, 26* (1979), 81–92.

Shackelford, Jole R. *Paracelsianism in Denmark and Norway in the 16th and 17th Centuries.* Madison: University of Wisconsin-Madison Doctoral Dissertation, 1989.

Shackelford, Jole R. "Tycho Brahe, Laboratory Design and the Aim of Science: Andreas Libavius versus Tycho Brahe." *Isis, 84* (1993), 211–30.

Shackelford, Jole R. "Early Reception of Paracelsian Theory: Severinus and Erastus." *Sixteenth Century Journal, 26* (1995), 123–35.

Shea, William R. "Descartes and the Rosicrucians." *Annali dell'Istituto e Museo di Storia della Scienza di Firenze, 4* (1979), 29–47.

Sherrington, C. S. *The Endeavour of Jean Fernal.* Cambridge University Press, 1946.

Solomon, Howard M. *Public Welfare, Science, and Propaganda in Seventeenth Century France: The Innovations of Théophraste Renaudot.* Princeton, New Jersey: Princeton University Press, 1972.

Stephen, Leslie. "William Coward." In *Dictionary of National Biography,* vol. 4, pp. 1298–99. Oxford University Press, 1921–1922.

Stigler, Stephen M. "Apollo Mathematicus: A Story of Resistence to Quantification in the Seventeenth Century." *Proceedings of the American Philosophical Society, 136* (1992), 93–126.

Sudhoff, Karl. *Bibliographica Paracelsica.* Berlin: Verlag Georg Reimer, 1894/reprinted Graz: Akademische Druck-u. Verlagsanstalt, 1958.

Szdlo, Zbigniew. "The Alchemy of Michael Sendivogius: His Central Nitre Theory." *Ambix, 40* (1993), 129–46.

Szdlo, Zbigniew. "The Influence of the Central Nitre Theory of Michael Sendivogius on the Chemical Philosophy of the Seventeenth Century." *Ambix, 43* (1996), 80–96.

Temkin, Owsei. *Galenism: Rise and Decline of a Medical Philosophy.* Ithaca and London: Cornell University Press, 1973.

Thackray, Arnold. *Atoms and Powers: An Essay on Newtonian Matter-Theory and the Development of Chemistry.* Cambridge: Harvard University Press, 1970.

Thomas, Sir Henry. "The Society of Chymical Physicians: an Echo of the Great Plague of London, 1665." In E. Ashworth Underwood, *Science, Medicine and History: Essays on the Evolution of Scientific Thought and Medical Practice written in honour of Charles Singer.* 2 vols. Oxford University Press, 1953, *2,* pp. 56–71.

Thorndike, Lynn. *A History of Magic and Experimental Science.* 8 vols. New York: Columbia University Press, 1923–58.

Valverde, José Luis. *La Farmacia y las Ciencias Farmaceuticas en la Obra de Suarez de Rivera.* Salamanca: University of Salamanca Press: Cuadernos de Historia de la Medicina Española, Monografias XIII, 1970.

Vandevelde. A. J. J. "Helmontiana." Five parts in *Verslagen en Mededeelingen. K. Vlaamsche Academie voor Taal-en Letterkunde.* Pt. 1 (1929), pp. 453–76; Pt. 2 (1929), pp. 715–37; Pt. 3 (1929), pp. 857–79; Pt. 4 (1932), pp. 109–22; Pt. 5 (1936), pp. 339–87.

Wall, Cecil. Abstracted and arranged from the manuscripts and notes of the late H. Charles Cameron. Revised, annotated, and edited by E. Ashworth Underwood. *A History of the Worshipful Society of Apothecaries of London,* Vol. 1, 1617–1815. London, New York and Toronto: Oxford University Press for The Wellcome Historical Medical Museum, 1963.

Walton, Michael T. and Phyllis J. "Witches, Jews, and Spagyrists: Blood Remedies and Blood Transfusion in the Sixteenth Century." *Cauda Pavonis, 15* (1996), 12–15.

Wear, A. R., French, K., and Lonie, I. M. *The Medical Renaissance of the Sixteenth Century.* Cambridge University Press, 1985.

Wolf, Abraham. *A History of Science, Technology, and Philosophy in the 16th & 17th Centuries.* New edition prepared by Douglas McKie. London: George Allen & Unwin Ltd., 1950.

Yates, Dame Frances A. *Giordano Bruno and the Hermetic Tradition.* Chicago: The University of Chicago Press, 1964.

Index